New Guinea:

Research from an Heterogeneous Island

Proceedings of the Biological Society of New Guinea
Meeting at Wau Ecology Institute, 1993

Edited by Robert Höft

New Guinea: Research from an Heterogeneous Island. Proceedings of the Biological Society of New Guinea Meeting at Wau Ecology Institute, 1993, edited by Robert Höft

Originally published by the Wau Ecology Institute as, "Proceedings of the Biological Society of New Guinea, Annual Meeting 1993". Reprinted with Permission.

ISBN 978-9980-945-93-8

Copies are available from:

University of Papua New Guinea Press and Bookshop
PO Box 413
University PO, NCD
Papua New Guinea

FOREWORD

The island of New Guinea, the second largest island in the world, has always intrigued me ever since I learnt about it in my school geography lessons.

Lying in the tropics straddling the nations of Papua New Guinea and Indonesia, the Island boasts one of the most diverse tropical flora and fauna in the world. The high level of biodiversity is presently undergoing strenuous pressures from human activities in the form of commercial and other resource development.

Unless studies are undertaken in every environment to determine the biodiversity, the subsequent resource development could well result in the extinction of species of fauna and flora without anyone ever knowing they existed. Particular plants or animals may easily be erased from the face of the earth.

The formation and existence of the New Guinea Biological Society must give the people of the two nations the only hope for the possible protection and preservation of indigenous fauna and flora of the Island of New Guinea. The efforts of those who had the foresight to initiate the society deserve praise from the two countries' governments and others concerned with the protection and preservation of the environment of New Guinea.

In this regard, it is encouraging to see the institutions of higher learning in Papua New Guinea and Irian Jaya taking interest in this field. Support agencies of the 1993 New Guinea Biological Society Conference including the University of Cenderawasih, CRA Minerals (PNG) Pty. Ltd., OK Tedi Mining Ltd., Papua New Guinea Forest Research Institute, University of Papua New Guinea, Papua New Guinea University of Technology, Papua New Guinea Forestry College, German Development Service and Wau Ecology Institute must see it as their objective to continue what has been started. It would be highly commendable if a permanent funding arrangement is in place for this purpose.

I hoped that this kind of support will continue to grow, so that in years to come our children and their children will have good reason to be proud of us as the custodians of Biology of New Guinea, which we only borrow from them.

Camillus S. Narokobi, LLB, LLM
Chairman
Wau Ecology Institute

PREFACE

This volume is a collection of abstracts, poster presentations, short communications and full papers reflecting the range of current research within the two nations on the Island of New Guinea.

It was felt by the majority of the participants that the papers presented at the meeting should be collected and published in a concise form. The result is as heterogenous as the Island. The papers have been accepted without refereeing. The editor tried his best to give them a uniform shape and work on language and spelling.

During the conference, botanists and entomologists from various places and institutions have come together and discussed their problems, needs, current and future projects. Three months after the meeting, the enthusiasm about possible co-operations and joint projects may have faded. Still, one fact should be remembered: both, botanists and entomologists have unanimously expressed the necessity to protect, maintain and eventually upgrade the national collections of plants and insects. A number of unspectacular institutions - the National Herbarium, the National Insect Collection and several smaller biological collections in the two countries, are the basis for all work - be it in the field of taxonomy, ecology, agronomy, forestry or conservation. Hardly any of the contributions included in this volume would have been completed without consulting one of these institutions. Hence, decision-makers were to be encouraged by all biologists working on the Island to support the national collections of plants and animals - as a matter of urge. This would be a major contribution to ensure successful meetings of the Biological Society of New Guinea in future.

Robert Höft

LIST OF CONTENTS

SESSION VII: SYSTEM ECOLOGY

SESSION VIII: CONSERVATION: CONCEPTS, PROBLEMS, NEEDS

SESSION I: VEGETATION SCIENCE

Proceedings of the Biological Society of New Guinea
R. Höft (ed.)

FOREST STRUCTURE OF A MONTANE RAIN FOREST IN WAU, MOROBE PROVINCE, PAPUA NEW GUINEA

Robert Kiapranis & Balun Lawong[†]

Papua New Guinea Forest Research Institute, P.O. Box 314, Lae, Papua New Guinea
[†] *Papua New Guinea Forestry College, P.O. Box 92, Bulolo, Papua New Guinea*

ABSTRACT

This paper presents a preliminary analysis of the forest structure of a montane rain forest studied in the Wau area, Morobe Province. The rain forest under investigation formed two distinct canopy layers and a shrub layer while the herbs formed the ground herb layer. The average tree height was about 24 m. The majority of the trees recorded were in the lower height classes (5-15 m and 16-25 m) as well as in the lower diameter at breast height (dbh) classes (10-20 cm, 21-30 cm and 31-40 cm). Very few trees with a diameter exceeding 70 cm were recorded in the plot.

INTRODUCTION

The montane rain forest in Papua New Guinea is still one of the least studied forest types in the world. Detailed information on the ecology, forest composition and diversity, as well as other ecological aspects is very limited and rather general. A number of reports have been published on general observations made on floristic composition and structure of montane forests in Papua New Guinea (Paijmans & Löffler 1972; Frodin 1977; Gressitt & Nadkarni 1978). The only detailed analysis from a montane forest was published by Hyndman & Menzies (1990) who worked around the headwaters of the Ok Tedi area. We are not aware of any detailed study on montane forests from the Wau area except for the field guide to Mt. Kaindi near Wau (Gressitt & Nadkarni 1978).

This paper presents part of a study aimed at providing a total plant species inventory from a montane rain forest area in Papua New Guinea. The paper gives details on the forest structure and composition.

STUDY AREA

The study was conducted in a montane rain forest mountain on the Kuper Range (146°43'E, 7°36'S), in the Wau District, Morobe Province, Papua New Guinea. The study site was established at an elevation of 1,992 m, about 30 km south-east of Wau township along the Wau-Biaru road. The plot was established on a gently sloping ridge below Mt. Kolorong.

The underlying geological structure of the Kuper Range is part of the New Guinea mobile belt which consists of low grade metamorphics and acid to basic and ultra basic plutons, most of which have been intruded along major fault planes (Löffler 1977). Detailed geological structures and features of the area have been published by Dow *et al.* (1974) in the Geological Series, Explanatory Notes of the Wau 1:250000 Geological Sheet, Papua New

Guinea.

A soil study of the area by Höft (in prep.) shows that the soils of the area consist of humic brown clay belonging to the Humitropepts group (USDA classification) which are derived from slates, phyllites and quartzic schists (Bleeker 1983). The thick, dark, topsoil is made up of slow decaying plant material. In the A-horizon, the soil consists of brown to dark brown clay while reddish clays are found below this layer.

The area is wet throughout the year. Rainfall data recorded from Edie Creek at a similar elevation for 25 years period show an average annual rainfall of 2737 mm (McAlpine *et al.* 1975).

General Features of the Forest Area

The major feature of the forest communities within the Kuper Range was the dominance of oak tree species *Castanopsis accuminatissima*, *Lithocarpus* sp. and *Nothofagus* spp. Other common tree species which were often very conspicuous in the forest area included *Planchonella montana*, *Elaeocarpus* spp., *Sloanea* spp. and *Schizomeria* spp.

On the forest floor, shrubby species of *Psychotria*, *Melastoma*, *Steganthera* and *Symplocos* were observed to be abundant. Several ginger species of *Alpinia* and *Riedelia* as well as the presence of bamboos were a common feature of the forest in the area.

Along the sides of the road, soon after the edge of grassland, climbing bamboos (*Nastus productus*) could be observed growing vigorously in the open areas. However, they seemed to be totally absent in other areas.

Methods

A field study survey of the area was carried out in early December of 1991 in which this study site was selected for the study. The forest area selected was considered to be representative of the vegetation type found in the area and easily accessible. An area of 60 m x 60 m (0.36 ha) was established and divided into 20 m x 20 m subplots (9 subplots). Each subplot was further divided into 5 m x 20 m rectangular plots (4 rectangular plots per subplot) and then each rectangular plot was again divided into 2 m x 5 m study plots (10 study plots per rectangular subplot). The smallest working unit area was a 2 m x 5 m rectangular plot area, which facilitated the total enumeration.

All trees exceeding 10 cm diameter at breast height (dbh) were measured and their heights estimated. Other plant life forms included in the survey were tree seedlings and saplings, shrubs, herbs, vegetative clumps (each clump was considered as one individual), epiphytes on host plants up to about 10 m high, and vines if they were rooted within the plot. Vouchers of the collections were processed and identified at the Papua New Guinea National Herbarium (Lae).

Results

Forest Structure

The forest structure of the montane rain forest studied shared general features reported from other montane rain forest areas (Richards 1964, Ohsawa *et. al.* 1985, Weaver 1990, Smiet 1992). The trees formed a tall closed canopy, except in areas where gaps had been created by tree fall. The forest consisted of two storeys, the shrub-layer and a layer of ground herbs. The crowns of the canopy trees tended to be narrow and compacted to some extent with very few vines and climbers. The dominant vines and climbers species recorded were

Frecyinetia spp. (Pandanaceae), *Piper* spp. (Piperaceae) and climbing *Psychotria* spp. (Rubiaceae). The branches of most of the upper canopy trees were observed to be covered with numerous epiphytes, particularly orchids.

The average tree height observed from the plot was about 24 m. About 83 percent of the trees (≥10 cm dbh) recorded were part of the second stratum of the canopy layer with heights ranging from 5 to 30 m while only 17 percent formed the uppermost canopy layer (Figure 1). The diameter class distribution of the trees is shown in Figure 2. As expected, there were more trees (94%) in the lower diameter classes (10-20 cm, 21-30 cm, 31-40 cm and 41-50 cm) while a similar number of individuals was recorded for each of the higher diameter classes.

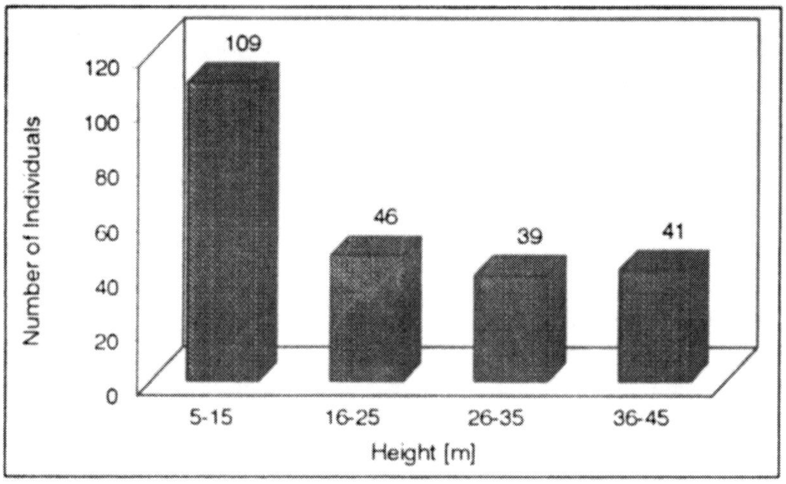

Figure 1. Tree height distribution in 0.36 ha plot in Kuper Range, Morobe Province

Figure 2. Diameter class distribution of tree species in a 0.36 ha plot in the Kuper Range

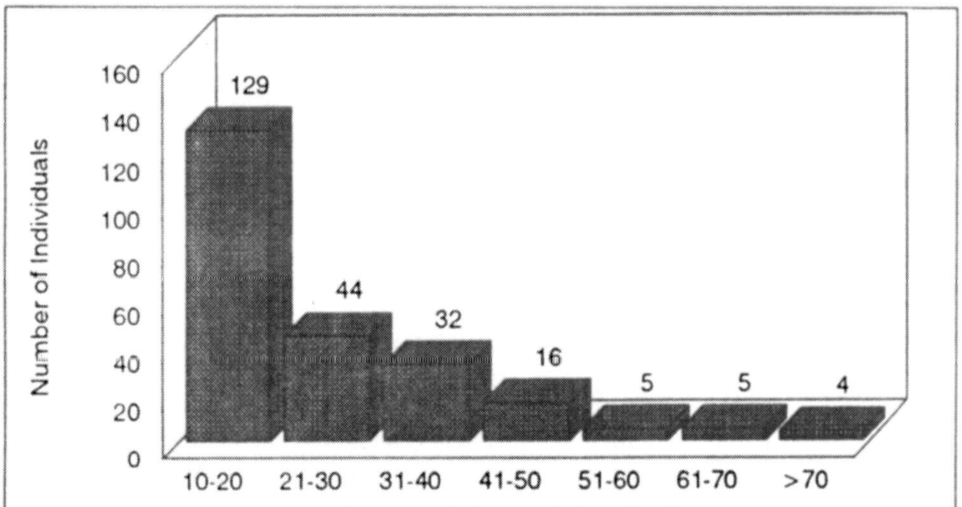

The most common trees were *Castanopsis accuminatissima* (Fagaceae) (78) and *Planchonella montana* (Sapotaceae) (22). Only one individual tree was recorded for each of the following species: *Fagraea* sp. (Loganiaceae), *Aglaia sapindina* (Meliaceae), *Myristica* sp. (Myristicaceae) *Finschia* sp. (Protiaceae) and *Saurauia* sp. (Saurauiaceae).

Shrubs

The shrubs which comprised woody plants of 1 to 5 m in height were observed to be generally abundant in the plot. The most abundant species belonged to the families Zingiberaceae, Melastomataceae and Gramineae (particularly bamboos). Palms which are quite common in lowland rain forests were less frequent except for *Calamus steenisii*, while tree ferns were more prevalent.

Herbs

The herbaceous plants were observed to be sparsely distributed throughout the plot. It was observed that the herbaceous species were more common in gaps but low in density under closed forest canopy areas. The small rubiaceous herb *Psychotria beccarii* was observed to be the most dominant species.

DISCUSSION

Detailed studies of the forest structure and composition as well as the general ecological characteristics of montane rain forests in Papua New Guinea are still very limited. The only detailed ecological analysis of a montane forest we are aware of was conducted by Gressitt & Nadkarni (1978) and Hyndman & Menzies (1990). Any comparison with other studies carried out in similar forests would not be possible at this stage because the results obtained here were from one single plot while most other studies focussed on vegetation changes in respect to altitude (e.g. Ohsawa *et al.* 1985; Hyndman & Menzies 1990; Weaver & Murphy 1990). However, based on information available on the general features and characteristics from similar forests (e.g. Richard 1964) we conclude that this forest area is structurally related to primary montane forests from other regions.

One of the features observed regarding the forest structure from the plot studied was that the number of individuals recorded in both the higher height and diameter classes were more or less similar. This may imply that only very few trees are capable of attaining the forest canopy and are able to occupy that niche for a longer period of time.

SUMMARY

The data presented here is only part (from 0.36 ha) of the data that has been collected for the proposed 1 ha study plot. To date, a total of 0.64 ha has been enumerated which may give a more detailed picture. Nonetheless, the data indicates that the Kuper Range montane rain forest is similar to other montane rain forests in terms of structure but more study is required to confirm these features of the forest area.

ACKNOWLEDGEMENTS

Without the help of many officers this work would not have been possible. We would like to thank the following officers for their invaluable support: Mr. Artis Vinas and Mr. Kenneth Mamu (Papua New Guinea Forestry College, Bulolo), Mr. Joe Wiakabu and Mr. Frank Asok (Papua New Guinea Forest Research Institute), Mr. Alberta Kairo (Wau Ecology Institute), Mr. Paul Katik and Mr. Kipiro Damas (Papua New Guinea National Herbarium, Lae) assisted in the identification of plant collections.

Mr. Balun Lawong allowed his 1992 second year students (Papua New Guinea Forestry College) to participate in part of the field work. We would like to thank Dr. Robert Höft (Wau Ecology Institute) for allowing us to use part of his data on soils and his suggestions. The Director of Wau Ecology Institute, Mr. Harry Sakulas, allowed us to use the Institute's field station facilities and its proposed conserved forest area for this study.

REFERENCES

Bleeker P. (1983). *Soils of Papua New Guinea.* Australian National University Press, Canberra, 352 pp.

Dow D.B., Smit J.A.J. & Page R.W. (1974). *1:250 000 Geological Series - Explanatory Notes. Wau, Papua New Guinea.* Australian Government Publishing Service, Canberra, 33 pp.

Gressitt, J.L., Nadkarni, N. (1978). *Guide to Mt. Kaindi: Background to Montane New Guinea Ecology.* Wau Ecology Institute Handbook No. 5, 135 pp.

Höft R. (in prep.). Soils of Mt. Kolorong Conservation Area, Kuper Range, Morobe Province, Papua New Guinea.

Hyndman D. C. & Menzies J.I. (1990). Rain forests of the Ok Tedi headwaters, New Guinea: an ecological analysis. *J. Biogeography* **17**, 241-273.

Löffler E. (1977). *Geomorphology of Papua New Guinea.* CSIRO and Australian University Press, 196 pp.

McAlpine J.R., Keig G. & Short K. (1975). *Climatic Tables for Papua New Guinea.* Division of Land Use Research Technical Paper No. 37, CSIRO Australia, 177 pp.

Ohsawa M., Nainggolan P.H.J., Tanaka N. & Anwar C. (1985). Altitudinal zonation of forest vegetation on Mt. Kerinci, Sumatra: with comparation to zonation in temperate region of east Asia. *J. Trop. Ecol.* **1**, 193-216.

Richards P.W. (1964). *The tropical rain forest. An ecological study.* Second edition, Cambridge University Press.

Weaver P.L, Murphy P.G. (1990). Forest structure and productivity in Puerto Rico's Luguillo Mountains. *Biotropica* **22(1)**, 69-82.

Proceedings of the Biological Society of New Guinea
R. Höft (ed.)

STUDY ON THE BIOLOGICAL DIVERSITY OF HUNSTEIN RANGE, EAST SEPIK PROVINCE, PAPUA NEW GUINEA

Lawong Balun, Artis Vinas & Larry Orsak

Papua New Guinea Forestry College, P.O. Box 92, Bulolo, Morobe Province, Papua New Guinea
Christensen Research Institute, P.O. Box 305, Madang, Papua New Guinea

ABSTRACT

The study of the biological diversity of the Hunstein Range was conducted at 50 m (site 1), 400 m (site 2), and 700 m (site 3) elevation on the northeastern side of Mt Samsai. The biological diversity was measured through direct and indirect sampling techniques.

A total of 146 plant species in 108 genera and 55 families was recorded on the 0.04 hectare (ha) plot (site 1, sandy alluvial rainforest). On site 2 (transitional forest type) a total of 155 species in 124 genera and 66 families was recorded in a 0.04 ha plot. At site 3 (lower montane forest) a total of 153 species in 119 genera and 61 families was recorded in the 0.04 ha plot. All species collected in this study had been previously recorded by Sohmer *et al.* in 1989.

An overall total of 233 species from 185 genera and 79 families were recorded in the study area. Among the 3 sites enumerated, 14 percent of the plant species were unique to a site. A cumulative species area curve shows that the species number *per se* did not level off after 0.12 ha.

The species diversity *per se* (233/0.12 ha) recorded from this study is one of the highest in the world. It is relatively very high compared to the highest species number *per se* (365/0.01 ha) recorded in Mexico by Gentry and Dodson in 1987. The diversity along the altitudinal gradation (beta diversity) increased from 50 m to 400 m but declined after 400 m. The decline in species number after 400 m is not significant.

From the total plant species recorded in this study, 5 percent are endemic. Flagship species such as Cassowary (*Casuarius bennetti*), Bird of Paradise (*Paradisaea* spp.), Bird wing Butterfly (*Ornithoptera goliath*), Vulturine Parrot (*Psittrichas fulgidus*) and Goura bird (*Goura cristata*) which are becoming rare throughout New Guinea were abundant in the Hunstein Range.

The luxuriant forest of Hunstein Range stocks on very shallow soil (10 - 30 cm thick). More than 80 percent of the forest occupies slopes and ridges which range from 30° to near 90° in certain instances.

Along the river banks, the forest floor is covered with successive layers of sediments. The forest floor due to frequent wash out from floods and successive deposition of sediments, was devoid of seedlings essential for maintaining the population.

Approximately, 18 percent of the total plant species recorded in this study are widely used in traditional medicine throughout Papua New Guinea. Commercial value of few of the insects collected or sighted ranged from 50 toea to K2 each (for buprestid beetles) and K30.00 each (for *Ornithoptera goliath*) at the current prices (Clark, pers. comm.).

INTRODUCTION

The Hunstein Range is a remote and little disturbed part of upper East Sepik Province. Encircled by the April and Salumei rivers, the Range rises above the Sepik River flood plain to over 1500 m altitude, and harbours a remarkably high diversity of flora and fauna. While there is some truth in claiming that the area has a very rich biodiversity there is no data to support. Nobody knows what kind of plants and animals and how many of them are there. Nobody even knows what proportion of the forest is traditionally utilized and what percentage of the local population utilizes the forest.

It is now generally accepted that every area of the organismic world has an unparalleled history and biological make up. The unparalleled history and biological composition of Hunstein Range area together with the very little data on biodiversity from Papua New Guinea makes it truly impossible to build any conservation models for the Hunstein Range from existing data sources. In terms of number of species *per se*, Kiapranis (1990) recorded a total of 363 species in a 0.64 ha lowland rain forest in the Gogol Valley, Papua New Guinea. Balun and Kiapranis (unpublished) recorded 351 species at 1,992 m elevation on a 0.36 hectare plot (see also Kiapranis & Balun, this volume). Paijmans (1970) in his study on four different rain forest community (0.8 ha) in New Guinea enumerated 122, 147, 145 and 116 species of trees above 30 cm girth.

Sohmer *et al.* (unpublished) in their 1989 Hunstein expedition recorded 231 genera in 112 families. It was estimated that about 40% of the collection was endemic to New Guinea. Sohmer and others recorded many vertebrate species, particularly birds and mammals, that have become rare or have gone extinct elsewhere. These were locally common in the Hunstein River Valley (e.g. Goura pigeon, Vulturine Parrot, Black Cockatoo).

Biodiversity is measured in several ways which can be divided into two major categories: direct sampling methods and indirect sampling methods.

- Direct sampling methods involve the enumeration (identifying species and measuring their abundance and distribution) of sample plots taken directly from recognizable forest types.
- Indirect sampling methods involve the application of diversity indices taken from sampling known indicator species (Holloway 1984).

Both sampling methods have their own bias and the side by side use of both methods is recommended.

Invertebrates, particularly insects, have recently been recognized to produce efficient and accurate measures of biodiversity (Herbert 1980; Ricklefs & O'Rourke 1984; Holloway 1984). Insects, such as moth and butterflies, have produced promising results for mapping patterns of species richness, areas of endemism, or taxonomic diversity (Miller & Holloway 1991).

The extent to which wildlife is utilized in the Papua New Guinea society varies from one place to another. Anthropologists, ethnozoologists, and biologists have documented that mans dependence on wildlife species for either subsistence or commercial purposes varies from one area to another, depending on the availability of wildlife in the area, accessibility of the area by road, the degree of economic development, and, most importantly, the socio-cultural activities being practiced by the people in the area (Bulmer 1982; Balun & Holdsworth 1988; Lamb 1990). Anyone coming up with a workable management strategy for a proposed conservation area (e.g. the Hunstein Range) should take into account the local villages or hamlets as an integral part of Hunstein Range during a rapid faunal and floral survey.

Thus, it is the main aim of this project to conduct a rapid biological survey of the Hunstein area in order to generate and furnish the resource managers with the necessary data for the development of the best management strategies for Hunstein Range (i.e. strategies for the conservation of natural life support systems such as protected areas should be compatible with the traditional rights and customs of the people, and implementation should be in consultation with and in agreement with the people involved in order to ensure successful conservation measures without unduly disrupting the traditional and customary practices).

This study, after completion, should provide the following data:

- Estimation of the number of plant species from subplots taken at regular intervals along a linear transect;
- Map of the biodiversity along altitudinal gradients in the Hunstein Range;
- Estimation of the number of plant species of socio-cultural value.

DESCRIPTION OF THE STUDY SITE

Location

The study was conducted in the Hunstein Range, upper basin of the Sepik plain, East Sepik Province, Papua New Guinea (04^0 20' - 04^0 32'S, 142^0 30- 142^0 46'E), at an altitudinal range of 10 m to 1544 m a.s.l. (Figure 1). Three different study sites were established at 50 m, 400 m, and 700 m above sea level on the northeast side of Mt. Samsai, the highest peak of the Hunstein Range (1540 m a.s.l). Site one was established at 50 m a.s.l (foot of Mt Samsai) the vegetation being classified as sandy alluvial forest type. The second study site was established at 400 m on the southeast side of Mt Samsai. The study site two displayed a lower montane forest type. The third study site was established at 700 m a.s.l. The forest type at 700 m a.s.l. was similar to site two in structural and floristic features.

Figure 1. Sketch map of the study area

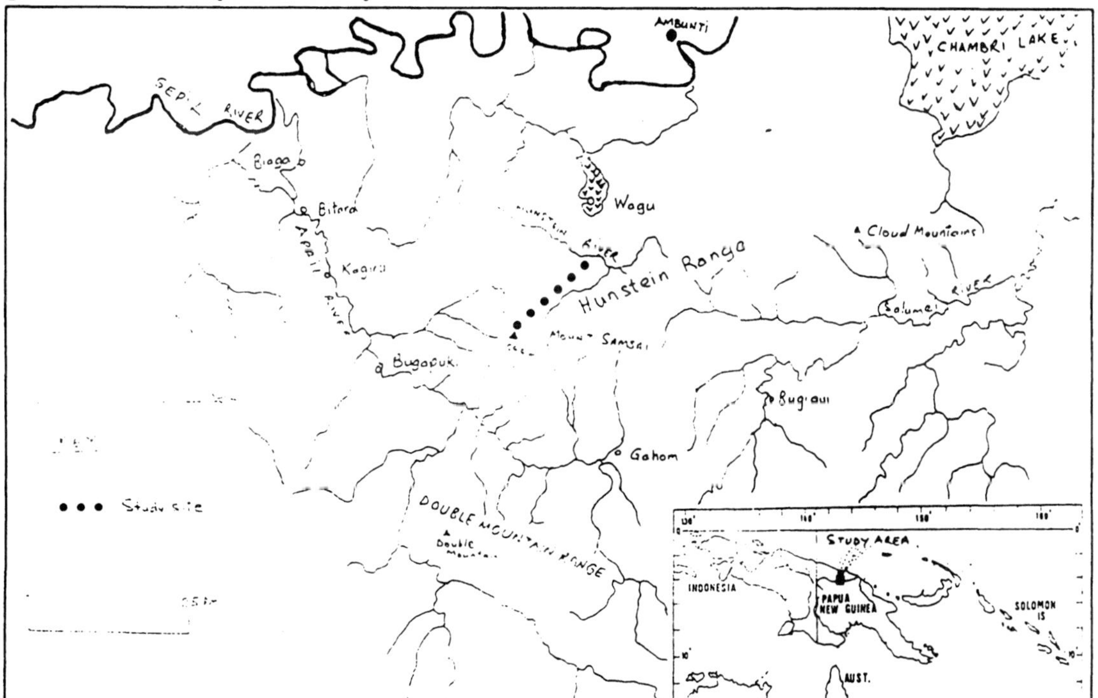

Topography

The Hunstein Range is encircled by low foothills and swampland and is drained by a number of fast flowing rivers (Figure 1). Geologically, the range is dominated by sedimentary rock types dissected by faults. The range is relatively unstable and has steep, rugged ridges which are prone to frequent land slips. The Hunstein Range forms a sudden rise of a mountain range from the typically low lying Sepik plains and swamplands. The valleys of the April and Salumei Rivers divide the Hunstein from the nearby highlands of Enga and Southern Highlands Provinces.

Soils

The soils from the different study plots were generally sandy alluviums at 10 - 50 m a.s.l and lateritic from 100 - 700 m a.s.l. The soils had a negligible humus layer in the sandy alluvium type and a 1 - 4 cm thick humus cover in the elevated zone. The soils from the foothills were devoid of humus layer due to frequent wash outs of the forest litter by run offs and high fluctuations in river water level. The soils generally had shallow profiles consisting of a thin (10-30 cm) solum with underlying parent material consisting largely of the sedimentary rock group.

Vegetation

The vegetation in all three study site was structurally, functionally, and physiognomically similar. Floristically, the study sites differed markedly in the dominance of species forming the canopy stratum. The forest at site 1 (alluvial montane forest type) had four distinct layers: emergent, main canopy, understorey, and shrub layer. The canopy structure of the forest at sites 2 and 3 were somewhat similar to the alluvial forest type.

Based on floristic features, the forests of the study sites could be classified into three forest-types or formations. The forest-type between 10-50 m a.s.l which stretched along the foot of Hunstein Range was sandy alluvial lowland rain forest. The second notable formation was lower montane forest which occupied the altitudinal span of 50 to 700 m a.s.l. The third recognizable formation was the mid-montane forest which occured above 700 m a.s.l.

Along the river banks, the forest floor was covered with successive layers of sediments. Due to frequent wash outs from floods and successive deposition of sediments the forest floor was largely devoid of vegetation.

Several other forest types bridging up with the study site were:

- Swamp forest: this formation formed the low lying (0 - 10 m a.s.l.) forest community. This forest type was characterized by an inundation period of no less than six months.
- Upper montane forest: this formation formed the forest community above the mid-montane forest type. This forest type was characterized by a noticeable reduction in forest structure and heavy loads of epiphytes.

Climate

Climatic data for Hunstein Range were not collected during our study. However, the rainfall and temperature data from the nearest station provided by Crittenden *et al.* (1988) could be referred to with some reliability.

Rainfall pattern is somewhat similar to the majority of the stations on the northern side of Papua New Guinea. A distinct wet/dry seasonality exists with nearly three quarters of the rain falling in the monsoon months of November to April. River water levels rise distinctly at this time. The dry season covers the remaining months of May to October. Mean temperatures are more or less constant throughout the year whereas the daily temperatures range between 23°C mean daily minimum and 32°C maximum. Winds are equally seasonal. Northwesterlies prevail in the monsoon period and southeasterlies during the dry season (Crittenden *et al.* 1988, in Chatterton & Waliawi 1991)

METHODS

Floristics

This study was conducted over 14 working days (11 January to 25 January 1993). The number of plant species and their patterns of distribution within the Hunstein Range along an altitudinal gradient, which is also termed as beta diversity (*sensu* McArthur 1965) was estimated analysing three sites at different elevations. The three sites were established at 50 m, 400 m, and 700 m a.s.l on the northeast side of Mt. Samsai, the highest peak of the Hunstein

Range (1540 m a.s.l).

The study site one was selected in a sandy alluvial forest type. This site displayed a forest with 4 distinct canopy strata; the forest floor was open. The second study site was established at 400 m up the southeast side of Mt Samsai. The study site two displayed a lower montane forest type. The forest floor of the second study site was moderately covered with seedlings and herbaceous species. The third study site was established at 700 m a.s.l. The forest type at 700 m a.s.l. was somewhat similar to site two in structural and floristic features.

At each study site, one 20 m x 20 m plot was established. The plots were subdivided into 10 m x 10 m subplots. Each 10 m x 10 m subplot was then further subdivided into 2 m x 5 m rectangles for easy accession.

One 2 m x 5 m plot at a time was assigned to the personnel trained on field identification of plant species for species enumeration. All plant species except bryophytes were enumerated. Seedlings and epiphytes up to 10 m above ground were also included. The positions of all trees with diameter at breast height (DBH) exceeding 10 cm were plotted on map sheets. Individuals with a DBH exceeding 10 cm were identified, measured and their height estimated.

Individuals of the same species were enumerated separately as seedlings, saplings, trees, shrub, herb, vines, climbers, vegetative clumps, and epiphytes.

All tree species with a DBH less than 10 cm and height between 1 and 5 m were recorded as saplings, while those less than 1 m height were recorded as seedling, shrub, or herb. Vines and climbers were counted only when they rooted within the plot. Individuals occurring as clumps were considered as one individual.

Voucher specimens of all sampled species were collected and pressed in the field. The pressed specimens were removed after one day and stored in 70% alcohol in air tight polythene bags. The specimens were then transferred to fresh newspaper, pressed and dried and identified at the Lae and Papua New Guinea Forestry College herbaria. Species lists and the species abundance of the three study sites were generated from the herbarium identification and field data. The identified specimens were mounted and stored at Papua New Guinea Forestry College herbarium.

Moth Collection

Approximately 8-9 nightly collections were made at study sites 1 and 2. Each non-rainy night, a gas pressure lamp was lit and a white linen was set up to attract moths. All moth specimens were collected in cyanide charged killing jars. On the following day, moth specimens were removed from the killing jars and stored between soft paper tissues in empty ice cream containers containing silica gel. Insects other than moths with hard wings, such as beetles, wasps, and bugs were collected in spirit.

Moth collections were sorted into morphospecies in Madang, using current species concepts as reflected in the largest Papua New Guinea moth collections at Canberra (CSIRO) and Honolulu, Hawaii (Bishop Museum). The numbers of each morphospecies was tabulated. These data were then run through a computer program to rapidly calculate various biodiversity indices (Magurran 1988). These included the Shannon Index, alpha index (log series), etc. Different indices reflect different assumptions that can be made about the samples, versus the overall populations.

The biodiversity indices, plus the number of species were used along with the plant data to map the patterns of biodiversity in the Hunstein Range.

Butterflies and Other Insects

General collecting of butterflies and other showy insects were casually made during the field study. This was to survey, in a preliminary way, the possible commercial utilization of Hunstein region insects as non timber forest products.

RESULTS

Floristics: Species composition and distribution

The plant species composition and abundance *per se* (diversity) was studied in three different sites. In site 1, a total of 144 species in 103 genera and 56 families was recorded in the 0.04 ha plot. In site 2, a total of 155 species in 118 genera and 65 families was recorded in the 0.04 ha plot. In site 3, a total of 149 species in 114 genera and 69 families was recorded in the 0.04 ha plot. Appendix 1 gives a species list. Species at 50 m a.s.l. (site 1) had 66.7 percent in similar with those at 400 m (site 2) and 700 m a.s.l. (site 3) while site 2 and 3 shared 92.86 percent. A total of 233 plant species in 185 genera and 79 families was recorded in this study.

Figure 2. Species/area curve for three 0.04 ha plots

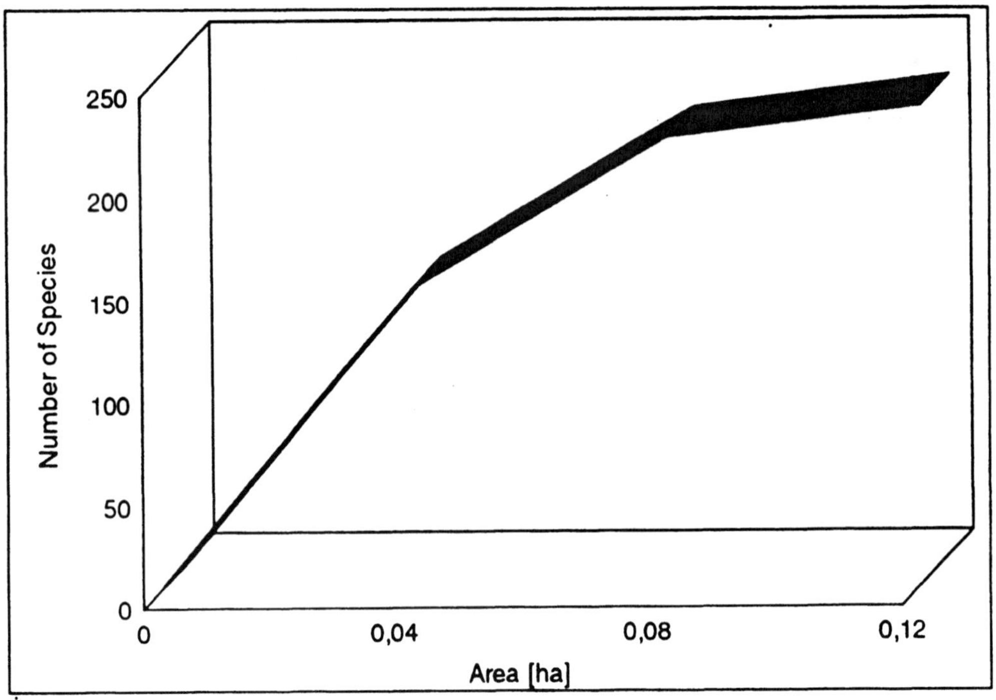

Endemism

Approximately 30 percent new species were recorded when the plot size was enlarged from 0.04 to 0.08 ha. When the plot size was increased from 0.08 to 0.12 ha, 14 percent new species were recorded (Figure 2). The cumulative species area curve indicates that the number of species *per se* does not level off after 0.12 ha. About 5 percent of the plant species recorded are endemic to New Guinea. Sohmer *et al.* (1991, unpublished) had recorded a rate of endemism of 40 percent.

Medicinal Value

Approximately 18 percent of the total plant species recorded in this study are treated in Balun & Holdsworth (1988). They are considered to be widely used in traditional medicine throughout Papua New Guinea. These medicinal plants are used traditionally for cure of diseases such as dysentery, sore eye, malaria, diarrhoea, cough, fatigue, stomach ache, wounds, as female contraceptives and many more.

Life Forms and survival

Table 1 gives the number of individuals representing different life forms at different elevations (study sites).

Table 1. Distribution of life forms (number of individuals and percentage in brackets) in three plots in the Hunstein Range

Major Life Forms	Site 1 (50 m)	Site 2 (400 m)	Site 3 (700 m)
Seedling	251 (25.61%)	222 (32.79%)	182 (21.36%)
Sapling	32 (3.25%)	42 (6.20%)	78 (9.16%)
Tree	10 (1.02%)	10 (1.48%)	6 (0.70%)
Herb/Shrub/Grass	248 (25.20%)	120 (17.73%)	175 (20.54%)
Epiphyte	309 (31.40%)	188 (27.77%)	297 (34.86%)
Vine/liana/creeper	122 (12.40%)	50 (7.39%)	67 (7.86%)
Palm	4 (0.41%)	40 (5.91%)	37 (4.34%)
Pandanus	7 (0.71%)	5 (0.74%)	10 (1.17%)
Total	984 (100%)	677 (100%)	52 (100%)

The predominant life form at sites 1 and 3 was the epiphyte while at site 2 seedlings dominated. In site 1, 12.67 percent of seedlings survived to reach the sapling stage and 3.97 percent became trees; in site 2, 18.92 percent of the seedlings survived to saplings and 4.51 percent to trees, and in site 3, 42.86 percent of the seedlings reached sapling form and 3.30 percent reached tree form (Table 1).

Forest Structure

Site 1 (sandy alluvial forest): This forest type occupied the bank of the Hunstein River and extended to the foot the Hunstein Range (100 m a.s.l.). The forest displayed three structural layers: main canopy stratum, middle stratum and shrub layer. The main canopy was characterized by the following species; *Dracontomelon dao*, *Spondias* sp., *Protium pinnate*, *Elaeocarpus longifolia*, *E. sphericus*, *Dysoxyllum* sp., *Maniltoa* sp., *Myristica* sp. and *Pometia pinnata*. The middle storey was dominated by *Polyalthia* sp., *Planchonella papuana*, *Chisocheton* sp., *Cryptocarya* sp., *Litsea* sp., *Palaquium* sp. and *Syzygium* spp. The main canopy and the middle storey were covered with climbers and stranglers including *Cissus* sp., *Lucinaea* sp., *Stephania* sp., *Tinospora* sp. *Derris* spp. and *Calamus* spp. Trees forming the main canopy had buttresses reaching up to 3 m height. The shrub layer was dominated by *Planchonella*, *Syzygium*, *Symplocos* and *Cryptocarya*.

The forest floor was sparsely covered by members of the family Zingiberaceae. The forest floor in site 1 was more or less devoid of proper seedlings.

Site 2 and 3 (lower montane forest): The forests at sites 2 and 3 were somewhat similar in their physiognomy. The leaves and buttresses were reduced in size. The forest was layered in three recognizable layers: main canopy, middle storey, and shrub layer. The main canopy was largely dominated by *Elaeocarpus* spp., *Planchonella* spp., *Myristica* sp. and *Pometia pinnata*. The middle storey was dominated by *Cryptocarya* sp., *Endiandra* sp., *Acmena* sp., and *Helicia* sp. The shrub layer was dominated by *Piper* sp., *Maranta* sp., and members of the family Zingiberaceae.

Moths

Relatively few moths were collected, due to the low intensity gas pressure lanterns used for the collection. Thus, moths collected on all dates at a given site were lumped for the biodiversity index calculations.

However, the normally high biodiversity expected in the tropics was replicated in the samples. Over half of the species collected were represented only by a single specimen.

By far the most common family in both samples were the Pyralidae.

Surprisingly few large moths (such as the Sphingidae) came to the light, apart from one night at Site 1. This paucity of large moths is most likely due to the type of light source. The collections in general were highly biased towards the family Pyralidae, which was most unusual, and again, probably due to the kind of lights used. In fact, very common families such as the Noctuidae and Geometridae were severely underrepresented in the sample. Only 6 species of green geometriines were collected (5 species from Site 1, 1 species from Site 2), and these were largely common species. Normally, in undisturbed forests, one would expect a dozen or more such green species.

One interesting collection was a large species of *Leuconota*. A 25% overlap in species between Sites 1 (110 species) and Site 2 (60 species) occurred.

Table 2. Biodiversity indices for the moth collection.

Site	1 (50 m a.s.l.)	2 (400 m a.s.l)
Number of species:	108	60
Number of individuals	285	136
% of total species captured represented by family Pyralidae	66%	83%
% of total individuals captured represented by family Pyralidae	52%	93%
Alpha biodiversity calculation	63.4%	41.1%
Alpha sum chi-square	6.744	14.1
Alpha, for Pyralidae only	30.94	29.9

Biodiversity indices for the moth collections (Table2) approximate that found in Madang Province lowland sites. Thus, species richness seems to reflect largely the characteristics of a typical lower elevation site in New Guinea. Sample sizes are too low to state with certainty the relative species richness at the two sites. The results of "alpha" are not statistically significant between the two sites, and when the moth family best represented in the collection is evaluated alone, "alphas" for the two sites are nearly identical. Pyralids are generally thought to be sensitive bioindicators, and since the type of light used so effectively attracted them, the data would be a good baseline for comparison.

Alpha seems to be the best biodiversity index to use for this data set, since it does not assume that the species collected in the sample reflects the actual species composition, as does e.g. Shannon Index.

Butterflies and Other Insects

Butterflies in the lowland forests and village area consisted of common species such as the Citrus-feeding *Papilio aegeus*, plus common *Taenaris* butterflies. These are generally widespread and not of much commercial value or of high demand. The papilionid (e.g., *Graphium aristeus*, *G. sarpedon*, and *G. eurypylus*) and nymphalid (e.g., *Hypolimnas bolina*) butterflies collected were all fairly common species, although the most interesting butterfly finding unfortunately could not be verified through collection of a voucher specimen.

A number of female birdwings (*Ornithoptera* spp.) were seen flying through undisturbed forest at about 600 meters. Most likely, these are specimens of *Ornithoptera goliath*, the world's second largest butterfly (Parsons 1991). The common birdwing species, *Ornithoptera priamus* and *Troides oblongomaculatus* simply do not occur in shady, undisturbed forests, nor was the *Aristolochia tagala* foodplants of those two species located during the survey. Another *Aristolochia* species, however, was found, and this could well be the foodplant of the sighted birdwing.

DISCUSSION

Papua New Guinea's lowland forests in general tend to have moderate biodiversity and low rates of endemism (that is, most of the species are widespread, extending even beyond the island of New Guinea). As altitude increases, endemism increases. This is probably true for biodiversity as well. There is much conflicting data as to which elevation is assiciated with the highest biodiversity. From qualitative surveys of weevils and other insects, Gressitt and Nadkarni (1978) felt, that the highest biodiversity occurred at around 1000-1200 metres. On the other hand, Allison *et al.* (see also this volume) in a series of fogging studies, found the highest biodiversity at 2000 metres, which is unexpected since areas above 2000 meters represent small "vegetational islands" and the theory of island biogeography, repeatedly tested, shows that smaller islands have fewer species. Work in South America indicates that the highest biodiversity might occur around 500 meters.

From a theoretical perspective, one would expect biodiversity in the Hunsteins to increase as one moves up the mountain from the lowlands, and topping out at some mid-elevation. Going further up than about 600 meters, the "vegetational island" areas become increasingly smaller, so one would expect biodiversity to decrease. On the other hand, there are probably low numbers of endemic plants and animals apart from the higher reaches of the Hunsteins. Due to the limited size of the higher elevation "vegetational islands" one would not expect endemic vertebrate species, apart from possible endemic rodents, frogs, and other fairly small species.

Plants: Species Composition, Abundance and distribution

The variation in species diversity between the three sites was not significant. However, the species composition between the three sites was noticeably different. For instance, of the total species recorded in the 400 m elevation, 30 percent of the species were new (not found in site 1 (50 m) elevation). About 14 percent of the species found in site 3 (700 m) elevation were unique to the site.

The results indicate that 5 percent of the plant species recorded may be endemic to New Guinea. About 18 percent of the total plants recorded are currently widely used in traditional medicine throughout Papua New Guinea.

The results of "alpha" diversity calculation for moths collected in sites 1 and 2 agrees with plant species diversity results, *i.e.* species diversity at sites 1 and 2 are identical. The data from site 1 compared to site 2 and 3 indicated a notable difference in plant species composition.

The presence of species like *Belliolum* sp. (Winteraceae), *Planchonella* sp. (Sapotaceae) and *Amaracarpus* sp. (Rubiaceae) indicate that the forest between 400 and 700 m elevation is a lower montane forest type or a transitional forest between lowland and lower montane forest. The data indicates that the sandy alluvial forest at 50 m elevation spreading at the foot of Hunstein Range is a distinct forest type while the forest at site 2 and 3 are of single type (lower montane forest type).

Plant diversity increased with altitude up to 400 m but declined between 400 and 700 m. Of the total 233 species recorded, 30 percent of the species at site 1 had no representatives in site 2. Approximately, 14 percent of the species recorded in site 3 were absent in site 1 or 2. The three data points used in this study only assume a vague pattern of spread in species diversity along an elevational gradient (beta diversity). Physiognomically, site 2 and 3 are similar. Hence, the difference in the number of species between the two sites may not go very far for any logical explanation in the trend of species diversity along an altitudinal gradient.

The Hunstein Range has a very rich and diverse flora by world standards. For example, the Hunstein species number (233/0.12 ha) is one of the highest recorded so far. The highest species number *per se* to date is 365/0.01 ha from Mexico (Gentry & Dodson 1987). Other high species diversity measured include; Whitmore's (1986) 236/0.01 ha from Costa Rica, which is the second highest. Gentry (1979); 151 species/0.1 ha from a wet tropical rain forest in the Panama canal zone, 125 species/0.1 ha from Panama again, and 139/0.1 ha recorded in Israel, Mediterranean zone by Naveh and Whittaker (1979).

Although the Hunstein data does not validate a direct comparison with similar studies done in Papua New Guinea because of different sampling techniques it can be speculated that on the basis of number of species *per se* when compared to 118 genera/0.4 ha. (Gressitt & Nadkarni, 1978), 364 species/0.64 ha. (Kiapranis, 1990) and 351 species/0.36 ha. (Balun & Kiapranis, unpublished) the Hunstein flora is relatively very rich.

In terms of altitudinal gradation of species richness, Diamond (in MacArthur 1972) in his diversity study of the avifauna of Mt. Karimui, in the highlands of New Guinea elucidated that the species diversity was highest at around 500 - 1000 m a.s.l but declined significantly and steadily after 1000 m a.s.l. Gressitt & Nadkarni (1978) indicated from their survey of Papua New Guinea beetles that the highest diversity is encountered at about 1200 metres.

Plants: Life Forms and their Patterns of Distribution

From the eight life forms recognized in this study, the predominant life form at site 1 and 3 was the epiphyte. The next dominant life form in sites 1 and 3 was seedling, followed by herb and vine. The predominant life form in site 2 was seedling followed by epiphyte and shrub, respectively. Results indicated that for the woody species which establish as seedlings under closed canopy the survival rate is very low (3.39%).

Moths

The general lack of geometrid moths, especially the green geometriine species, which are exceptionally good indicators of undisturbed primary forests, indicates that the lights used were not sufficient to draw in species from the upper canopy. Pyralid moths, which are heavily represented in the collections, tend to occur near the ground. Due to logistical problems, mercury vapor lamps and ultraviolet light setups could not be used. The data can therefore not be compared with other sites from Papua New Guinea. However, the richness of the pyralids that were collected indicates that the insect fauna is probably as rich as any lowland site.

No endemic species were detected in the samples, although for some Papua New Guinea moth groups, there are numerous undescribed species whose geographical distribution is yet unknown.

Butterflies and Other Insects

In Papua New Guinea, the more valuable insects from a commercial point of view (*i.e.* to be sold to the Bulolo Insect Farming and Trading Agency) are generally found in forested areas. We collected a number of iridescent day-flying scarab and buprestid beetles. Clark (pers. comm.) estimated the value of some forest-dwelling beetle species at 50 toea to K2 each.

Possibly the most valuable finding was the sighting of what is probably the world's second largest butterfly, *Ornithoptera goliath*. One village in Morobe Province is currently "ranching" this species, under the direction of the Insect Farming & Trading Agency; those specimens are being sold at K30 each. Lowland sites tend to have larger individuals of this species (Haugum & Low 1978) which makes them more valuable.

The foodplant of *O. goliath* in the Sepik area is a species tentatively known as Aristolochia crassinervia (Parsons 1983). The possibility of this butterfly being ranched must be weighed against the time involvement in setting up such a farm properly. Ranching the protected, primary forest-inhabiting birdwings, is not a matter of just planting the right caterpillar foodplant. These birdwings are highly localized because they are specialized, and those specializations must be addressed in a butterfly ranching program.

Ecological Implications

Hunstein Range has a high faunal and floral diversity (Sohmer *et al.* 1991, unpublished). The great diversity of wildlife, including flagship species like; Cassowary (*Casuarius bennetti*),

Bird of Paradise (*Paradisaea* spp), Birdwing Butterfly (*Ornithoptera goliath*), Vulturine Parrot (*Psittrichas fulgidus*), megapod bird and Goura bird (*Goura cristata*) in the Hunstein are highly habitat-specific. The diverse and luxuriant forest that harbours and supports the rich fauna is stocking on a very thin layer of top soil underlain by the parent rock.

The highly habitat-specific nature of these animals and the thin layer of soil that support the great plant biomass implies that the Hunstein Range is a very fragile and complex ecosystem. Unraveling this complexity makes us feel even less knowledgeable about the functioning of the natural systems that we utilize.

The need to identify the species that are important parts of the ecosystem, that form structure, enhance productivity, and recycle essential nutrients in the forest ecosystem is far from being attended. We know pitifully very few species as indicators of ecological processes or as keystone species that influence community structure as is the case of Hunstein Range. All these are definite warnings that any major disturbance in the Hunstein Range would entail serious repercussions including species extinction and genetic degradation.

The mere inconsistency between the data on species number published so far for Papua New Guinea may cause one to say, "only the future will tell how many species there are and which environments are most diverse". The flagship species found in the Hunstein Range are ecologically distinct populations which to some extent is the precondition for species endemism. The Hunstein Range may stand out as one of the biodiversity "hot spots" in Papua New Guinea with an exceptional level of endemism. Such areas are known to be facing greatest threats finally leading to species extinction. Conversion of the natural forest communities in the Hunstein Range could well eliminate large numbers of species. Of course, we may learn how to manipulate habitats to enhance survival prospects.

We may learn how to propagate threatened species in captivity. We may be able to apply emergent techniques, all of which could help to relieve the adverse repercussions of broad-scale deforestation. But at this stage, the damage will already have been done. There is no substitute for species extinction and speciation processes in artificial species enrichment programmes. As in the graphic phrasing of Soulé and Wilcox (1980), "Death is one thing; and end to birth is something else". Large forested areas (over 80 percent) of the Hunstein are stretched over slopes and ridges with 30° to near 90° inclination. Conversion of this forested area for large scale timber production would be a breach of the country's forestry regulations. The remaining 15 to 20 percent of the forest in the Hunstein is an alluvial lowland rain forest. This forest type is seasonally flooded with fluctuations in the water level. Much of the forest floor in the alluvial forest was devoid of vegetation, seedlings in particular.

The very little number or absence of seedlings on the forest floor warns that careless extraction of large tree species would be catastrophic to regeneration. The removal of forest would lead to serious erosion of the slopes, ridges and river banks. Erosion of river banks, increase in water turbidity and sedimentation would no doubt pose a major threat to the ichthyo-fauna (local fish species). The bark of *Pometia pinnata* (a dominant tree species) contains saponins. If these saponins are allowed to leach into streams and waterways, death of fish and crustaceans will be imminent.

Fresh water turtles and some important fish species which contribute a substantial amount of protein use sand banks along the river to lay eggs. Thus, erosion of river banks will be a major threat to these species.

Socio-Cultural Implications

Forest ecosystems are nature's warehouses, containing a variety of products. White (1976) pointed out the importance of forest products for Papua New Guinea's village life. Forests offer food (fruit, leaves, tubers), spoils of hunting and trapping (birds decoration, clothing, building and thatching materials, lianas (ropes), weapons, and poisons. A survey on diets of Wopkaimin people in Kiunga showed that the villagers obtained 69 kg of wild vegetables, 190 kg of pigs and 156 kg of wild animals in one month from the primary forest

(OTES, 1982). Liem (1975), Kwapena and others who studied the amount of wildlife products harvested in Papua New Guinea villages indicated that villagers at Garu village, Talesia, West New Britain Province had harvested 24,129 megapode eggs and 130 wild pigs from the forest between August 1973 and July 1974 (Liem, 1975).

People from Kanabea and Kaintiba in the Gulf Province utilized 21 species of plant from primary forest for all their ailments and extracted traditional salt from the ashes of three plant species (Balun & Holdsworth 1988). A pilot survey on the traditional medicinal plants of Morobe Province carried out by Wau Ecology Institute between 1985 and 1987 indicated that all the villages surveyed continued to utilize medicinal plants for most of their physical ailments. The villages and hamlets in the Hunstein Range no doubt use traditional medicinal plants for much of their medical needs. The strong dependency on forest products by the Hunstein populace for their subsistence needs warns that large scale disturbance of natural habitat would have serious socio-cultural effects.

Potentials for Natural Forests to Improve Quality of Life: Potential for New Drugs

Of the estimated 250,000 to 750,000 higher plants described to date on the surface of the earth only a handful have been exhaustively studied for their potential value as a source of drugs. It is interesting to know, that the 119 plant-derived drugs in use throughout the world today are obtained from less than 90 species of plants (Farnsworth *et al.* 1985). Of the 119 plant-drugs, 74 percent were discovered because of medicinal folkloric information. It is interesting that *Alstonia scholaris* (Apocynaceae) used throughout Papua New Guinea including Sepik area for treating malaria is the same species from which quinine (a currently prescribed antimalarial drug) was first discovered.

Science has proved, that success of research into new drugs is dependent on two main building blocks; availability of medicinal folkloric information and the availability of wild plant species. Removal of natural forest in the Hunstein Range would result in the loss of the natural source of traditional medicine and the potential bank for discoveries of new drugs. The benefits and the prospects of tapping from natural forest is of incalculable value. However, the prospects of natural forest yielding new products of commercial value rest on our ability and responsibility to conserve biological diversity.

Potentials for Natural Forests to Improve Quality of Life: New and Improved Crops

Tropical forest plants are known to contribute to agriculture in three different aspects:

- as sources of new crops that can be cultivated;
- as source material for breeding improved plant varieties; and
- as sources of new biodegradable pesticides.

Of the several thousand species known to be edible, only about 150 have ever become important enough to enter world commerce (Plotkin 1988). Today, less than 20 plant species produce most of the world's food (Vietmeyer 1986). It is interesting to know that much of our diet today has been determined by exploitation of the dietary patterns of the tropical countries. Many currently underexploited tropical species stigmatized as slave foods by the western colonials have the potential for yielding new crops. Many of which often tend to be robust, productive, self-reliant, free of indigestible compounds with relatively high nutritive value, and suitable for growing in some sort of agricultural system.

Improvement of crop species through cross-breeding: Relatives of commercial species must continuously be crossbred with these species to improve crop yield, nutritional quality, durability, responsiveness to different soils and climates, and resistance to pests and diseases (IUCN 1980). Since many of the world's most important crop species originated in the tropics, man must look to the equatorial regions for wild types to maintain and improve crop species. For instance, coffee (*Coffea* spp.) which is one of the major commercial crops in Papua New Guinea is susceptible to certain fungal diseases. Although Africa is the origin of *Coffea arabica*, the island of Madagascar has approximately 50 wild species of *Coffea*.

Some of these species may stand out as an important material for breeding programs leading to increased resistance to fungal infections and may also produce beans with little or no caffeine (Plotkin 1988). One classic example of wild type back-cross leading to improved quality of crop is the discovery of two new species of tomato in 1976 from a 1962 collection. In 1980 the back-crossing in one of the species yielded new tomato strains with larger fruit and increased content of soluble solids which boosted the U.S. economy by tens of millions of dollars (Iltis 1988).

Potentials for Natural Forests to Improve Quality of Life: Natural Pesticides

Many tropical plants have developed chemical defenses to deter predation by herbivorous animals. Tropical people possess a sophisticated knowledge of these plants, often using them as medicines or poisons. The calabar bean (*Physostigma venenosum*) was traditionally used as an ordeal poison in West Africa, and studies of the active principle of this species led to the development of methyl-carbamate insecticides.

World trade in daisy flowers (*Chysanthemum cinerariaefolium*), the source of insecticidal pyrethrum extracts, is a multimillion dollar business (Plotkin 1988). This plant was first discovered because of its use by an African tribe to control pests.

These are just a few examples of plant species which have contributed immensely and have the potential to improve our quality of life. For this reasons, among others, I have no patience with the phony requests of developers, economists, and humanitarians who want us biologists to "prove" with hard evidence, right here and now, the "value" of biodiversity and the "harm" of tropical deforestation. Rather, it should be for them, the sponsors of reckless destruction, to prove to the world that a plant or animal species, or an exotic ecosystem, is not useful and not ecologically significant before permitting society to destroy it. Such a proof, of course, can be offered neither by them nor anybody else!

Recommendations

The Hunstein Range has a relatively very rich fauna and flora. The area according to Sohmer *et al.* (1991) harbours 40 percent endemic species while this study recorded a rate of endemism of 5 percent. Hunstein Range, in these respect, may well and truly be one of the biodiversity "hot spots" in Papua New Guinea, hence, deserves to be conserved has a natural heritage area.

Of the total species enumerated in this study, 18 percent are widely use in traditional medicine throughout Papua New Guinea. The Hunstein villages are very remote from the nearest Government hospital (Ambunti). A study looking into possibilities of promoting the utilization of medicinal plants in the Hunstein villages is recommended.

Over 80 percent of the forest area in the Hunstein Range is spread over slopes and ridges with gradients between 20 to near 90° in same instances. The luxuriant forests are encountered stocking on a thin layer of soil. The alluvial forest type in the low lying areas along the river banks experience seasonal flooding and inundation.

The forest floors along the river banks consequently are deprived of significant accumulation of organic litterfall and seed propagules from the standing phytomass. These physical aspects of the Hunstein Range compounded with the biological complexities through which the forest tree species maintain their population discourage large scale commercial logging because of the potential risks to environmental pollution.

The cumulative species area curve suggests that the number of species beyond 0.12 ha is still on a steady rise. Hence, a further work considering the enumeration of a larger plot size at regular altitudinal intervals is strongly recommended.

The number of species at 400 m and 700 m a.s.l are not markedly different despite a slight drop in number of species at 700 m a.s.l. Hence, a further study reaching the highest peak of Hunstein Range is desirable to establish the elevation range which supports the maximum number of species in the Hunstein Range.

The Hunstein river and the tributaries support a rich ichthyo-fauna (fish, prawns, and turtles) which local people depend on. Exploitation of the forest by large scale commercial logging entails the potential to generate pollution from saponins, chemicals, diesel fuel, oils and increase the load of sediments in the water body. The following approaches would provide equivalent or even better and sound development to the local population:

- Use of *wokabaut* sawmills to selectively extract timber in order to reduce the intensity of erosion and prevent saponins from entering the waterways. The activities of a *wokabaut* sawmill may prove as disastrous as large-scale logging if not supervised by properly trained personnel.
- The currently operating *wokabaut* sawmills in the area seem to lack appropriate supervision and monitoring. These problems must be curtailed immediately to ensure sustainable operation with the *wokabaut* sawmill.
- The impact of *wokabaut* sawmills on the environment is influenced by the responses from the local abiotic and biotic factors. These factors vary from place to place even within the same geographical region. Hence, an immediate research on the impact of *wokabaut* sawmills in the Hunstein area is strongly recommended to develop a relevant strategy.

ACKNOWLEDGEMENTS

The Australian Foundation for the Peoples of the South Pacific provided the funds for the field study in the Hunstein Range between January and February 1993. The East Sepik Council of Women assisted with the logistics during the field study. Ms. Mary Soondrawu, personally involved in organizing with villages and the women group in Maprik, Ambunti and Wagu area, assisted the research team throughout the period of study. Christensen Research Institute provided accommodation, laboratory space and access to telephone, fax and computer facilities during the transit.

The Papua New Guinea Forestry College facilitated transport from Bulolo to Madang and provided most of the camping and field equipment used in the project.

The following students assisted in the field survey: Mr. Alex Manui, Mr. Anton Galus Wasukware, Mr. Philip Topaiman, Mr. Francis Gumbali, Mr. John Wanam, Mr. Alois Jenkin and Mr. Peter Maken. The people from Wagu village, in particular Mr. Lukas Ketapi and family and Mr. Topaiman assisted greatly in the logistics in the field.

REFERENCES

Balun, L. & Holdsworth, D. (1988). Ethnomedicine of the Gulf Province of Papua New Guinea Part I: Mountains around Kanabea and Kaintiba. *Int. J. Crude Drug Res.* **26(1),** 51-55.

Bulmer, S. (1982). Human ecology and cultural variation in prehistoric New Guinea. *Biogeography and Ecology of New Guinea* (Ed. by J.L. Gressitt) Dr W. Junk, The Hague.

Crittenden, R. et al. (1988). Gavien and Kaupena: Assessing the Impacts of Development Projects in the East Sepik and South Highlands Province. *Dept. of Geography University of Papua New Guinea Occassional Paper* No **8,** Port Moresby.

Diamond, J.M. (1972). The spectrum of resources in the tropics. *Geographical Ecology: Patterns in the species distribution.* (Ed. byH.R. MacArthur), pp. 205-213, Harper Row Publ., Inc., N.Y.

Daly, H. V., Doyen, T.J., & Ehrlich, R.P. (1978). *An Introduction to Insect Biology and Diversity.* McGraw-Hill, New York.

Farnsworth, N.R., Akerele, O., Bingel, S.A., Soejarto, D.D., and Guo, Z.G. (1985). Medicinal plants in therapy. *Bull. WHO* **63,** 965-981.

Gentry, A.H. (1979). Extinction and conservation of plant species in Tropical America: A phytogeographical perspective. *Systematic Botany, Plant Utilization and Biosphere Conservation.* (Ed. by Hedberg, I.), pp. 110-126, Almqvist and Wikell International, Stockholm.

Gentry, A.H. & Dodson, C. (1987). Contribution of nontrees to species richness of a tropical rain forest. *Biotropica* **19(2),** 149-156.

Gressitt, J. L. & Nadkarni, N. (1978). *Guide to Mt. Kaindi. Background to Montane New Guinea Ecology.* Wau Eclogy Institute Handbook No. 5, 135 pp.

Haugum, J. & Low, M.A. (1978). *A Monograph of the Birdwing Butterflies. The Systematics of Ornithoptera, Troides and Related Genera. Vol 1. The genus Ornithoptera.* Scandinavian Science Press, Lt., Klampenborg, Denmark.

Hebert, P.D.N. (1980). Moth communities in montane Papua New Guinea. *Journal of Animal Ecology* **49,** 593-602.

Holloway, J.D. (1984). Moths as indicator organisms for categorising rain-forest and monitoring changes and regeneration processes. *Tropical Rain Forest: The Leeds Symposium* (Ed by A.C. Chadwick & L. Sutton), Leeds Philosophical & Literacy Society, Leeds, United Kingdom.

Iltis, H.H. (1988). Serendipity in the exploration of Biodiversity. What good are weedy tomatoes? *Biodiversity* (Ed. by E.O.Wilson), pp. 98-104, National Academic Press, Washington D.C.

IUCN (1980). *World Conservation strategy.* International Union for the Conservation of Nature, Gland, Switzerland, 55 pp.

Kiapranis, R. (1990). Plant species enumeration in lowland rain forest in Papua New Guinea. *Proceedings of International Workshop held on 19-29th November 1990.* (Ed. by D.A. Taylor & K.G. MacDican) Winrock Int. Inst. Agric., Los Baños.

Lamb, D. (1990). *Exploiting the Tropical Rain Forest: An Account of Pulpwood Logging in Papua New Guinea.* Man and the Biosphere Series Volume 3. Unesco, Paris.

Liem, D.S. (1975). Wildlife utilization in the proposed Garu wildlife management area, West New Britain, Papua New Guinea. Paper presented to the 1975 Waigani Seminar, Port Moresby, May 1976.

MacArthur, H.R. (1965). Patterns of species diversity. *Biol. Rev.*: 510-533.

MacArthur, H.R. (1972). *Geographical Ecology: Patterns in the species distribution.* Harper & Row Publ. Inc. N.Y.

Magurran, A.E. (1988). *Ecological Diversity and Its Measurement.* Princeton University Press, Princeton, New Jersey. 179 pp.

Miller, E.S. & Holloway, J.D. (1991). Priorities for Conservation in Papua New Guinea: Non-Marine Invertebrates. *Conservation and Environment in Papua New Guinea: Establishing Research Priorities. Proceedings of a Symposium* (Ed. by M. Pearl, B. Beehler, A. Allison, & M. Taylor), WCI, a division of New York Zoological Society, New York, U.S.A.

Naveh, Z. & Whittaker, H.R. (1979). Structutal and floristic diversity of shrublands and woodlands in northern Israel and other Mediterranenan areas. *Vegetatio* **41,** 171-190.

Noss, R. F. (1990). Indicators for monitoring biodiversity: a hierarchical approach. *Conservation Biology* **4,** 355-364.

OTES (Ok Tedi Environmental Study) **(1982).** *Ok Tedi Environmental Study main report.* By Maunsell & Partners PTY LTD Vol 1.

Paijmans, K. (1970). An analysis of four tropical rain forest sites in New Guinea. *J. Ecol.* **58,** 77-101.

Paijmans, K. (1976). *Vegetation in New Guinea.* Ed. by K. Paijmans, CSIRO, Australia, pp. 23-105.

Parsons, M. (1983). *A conservation study of the birdwing butterflies, Ornithoptera and Troides (Lepidoptera: Papilionidae) in Papua New Guinea.* Final report to the Department of Primary Industry, Papua New Guinea. 111 pp.

Plotkin, J.M. (1988). The outlook for new agricultural and industrial products from the tropics. *Biodiversity* (Ed. by E.O.Wilson), pp. 106-116, National Academic Press, Washington D.C.

Parsons, M. (1991). *Butterflies of the Wau-Bulolo Valley.* Bishop Museum Press, Honolulu, Wau Ecology Institute Handbook No. 12.

Ricklefs, R.E. & O'Rourke, K. (1974). Aspect of diversity: a temperate-tropical comparison. Evolution.

Sohmer, S.H., Kiapranis, R., Allison, A, Takeuchi, W. (1991, unpublished). Report on the Hunstein River Expedition -1989.

Vietmeyer, N. (1986). Lesser-known plants of potential use in agricultural and forestry. *Science* **232,** 1379-1384.

White, K.J. (1976). *Planned developments in lowland rain forest in Papua New Guinea with special reference to infrastracture, mining and forestry.* Report of MAB Symposium, April 1975, Port Moresby. MAB Publ. No. 3

Whitmore, T.C. (1986). Total species count on a small area of lowland tropical rain forest in Costa Rica. *Bull. Br. Ecol. Soc.* **17,** 147-149.

Appendix 1

Checklist of the plants from three plots in the Hunstein Range (Collection numbers in brackets)

Acanthaceae	Acanthus sp (B190)
Actinidiaceae	Saurauia sp (S3086)
Anacardiaceae	Buchanania sp (S3212)
Anacardiaceae	Dracontomelon dao (B355)
Anacardiaceae	Semercarpus sp (S3201)
Annonaceae	Polyalthia sp. (S3145)
Annonaceae	Popowia sp (S3179)
Apocynaceae	Parsonsia currisepala (S3039)
Araceae	Alcasis sp1 (B142)
Araceae	Alocasia sp (B015)
Araceae	Epipremnum sp (S3148)
Araceae	Holochlamys becarii (S3011)
Araceae	Phothos rumphii (S3233)
Araceae	Pothos albertisii (B092)
Araceae	Rhaphidophora geniculata (S3222)
Araceae	Scindapsus cf. novoguineensis (B010)
Araliaceae	Osmoxylon micranthum (S3200)
Araliaceae	Schefflera sp1 (B780)
Araucariaceae	Agathis labillardieri (S3135)
Arecaceae	Calyptocalyx clemensae Burrett (S3002)
Arecaceae	Calamus sp1 (S3169)
Arecaceae	Calamus sp2 (S3170)
Arecaceae	Orania sp (B384)
Arecaceae	Ptychosperma lautebachii (S3119)
Aristolochiaceae	Aristolochia cf. momandul (S3142)
Asclepiadaceae	Lamariopsis kingii (K3143)
Asclepiadaceae	Marsdenia sp (S3070)
Asclepiadaceae	Secomone sp (K3036)
Aspleniaceae	Asplenium sp1 (S3080)
Aspleniaceae	Asplenium sp2 (S3009)
Aspleniaceae	Asplenium sp3 (S3220)
Aspleniaceae	Asplenium sp4 (S3145)
Aspleniaceae	Diplazium weinlandii (B086)
Aspleniaceae	Lamariopsis kingii (S3143)
Aspleniaceae	Lecanopteris sirosa (B051)
Barringtoniaceae	Barringtonia sp (B716)
Barringtoniaceae	Planchonia papuana (B161)
Begoniaceae	Begonia sp. (S3148)
Burseraceae	Canarium sp (S3132)
Burseraceae	Protium sp (B721)
Burseraceae	Protium sp1 (B079)
Celestraceae	Sulucia sp (S3156b)
Clusiaceae	Calophyllum saulaterii (B031)
Clusiaceae	Garcinia dulcis (S3191)
Clusiaceae	Garcinia solomonensis (S3089)
Clusiaceae	Garcinia sp (S3204)
Combretaceae	Terminalia kaernbachii (K3218)
Combretaceae	Terminalia macrocarpa (K3136b)
Combretaceae	Terminalia sp (B0100)
Convolvulaceae	Merremia peltata (B382)
Cucurbitaceae	Mikania sp (S3159)
Cucurbitaceae	Momodica charantia (S3045)
Cunoniaceae	Caldcluvia sp1 (S3217)
Cyatheaceae	Cyathea sp (B738)
Davalliaceae	Humata cf. Similis (S3075)
Davalliaceae	Humata sp (S3088a)
Davalliaceae	Humata sp1 (S3004)
Davalliaceae	Humata sp2 (S3008)
Davalliaceae	Nephrolepis cordiflora (B953)
Davalliaceae	Nephrolepis sp (B689)
Dioscoreaceae	Dioscorea sp (B968)
Dipterocarpaceae	Anisoptera polyandra (S3213)
Dipterocarpaceae	Vatica papuana (S3025)
Dracaenaceae	Dracaena sp (S3196)
Elaeocarpaceae	Elaeocarpus ledermannii (S3143)
Elaeocarpaceae	Elaeocarpus longifolum (S3254)
Elaeocarpaceae	Elaeocarpus sp (S3068)
Elaeocarpaceae	Sloanea tigii (S3199)
Elaeocarpaceae	Solanea sogerensis (S3132)
Ericaceae	Dimorphanthera sp (S3065)
Euphorbiaceae	Aporusa papuana (B035b)
Euphorbiaceae	Codiaeum sp (B016)
Euphorbiaceae	Drypetes sp (S3076)
Euphorbiaceae	Glochidion sp1 (B791)
Euphorbiaceae	Macaranga sp1 (B857)
Euphorbiaceae	Macaranga sp (S3115)
Euphorbiaceae	Mallotus philippinensis (B770)
Fabaceae	Albizia falcateria (B209)
Fabaceae	Archidendron glabrium (S3158)
Fabaceae	Derris pinnata (B386)
Fabaceae	Derris sp (B700)
Fabaceae	Instia sp (K3072h)
Fabaceae	Kingiodendron sp (K3047)
Fabaceae	Maniltoa sp (B778)
Fabaceae	Pithecolebium sp (B126)
Flacourtiaceae	Flacourtia sp (B359)
Flacourtiaceae	Osmelia philippina (S3058)
Flacourtiaceae	Pangium edule (B043)
Flagellariaceae	Flagellaria indica (S3012)
Gesneraceae	Cyrtandra sp1 (S3153)
Gesneriaceae	Cyrtandra sp (S3120)
Gnetaceae	Gnetum gnemon (S3211)
Hymenophyllacea	Callistopteris apiifolia (S3133)
Hymenophyllaceae	Sphaerocionium sp. (S3228)
Icacinaceae	Gomphandra montana (S3140)
Icacinaceae	Gonocaryum litorale (B034)
Icacinaceae	Medusanthera laxiflora (S3043)
Icacinaceae	Rhyticaryum sp (B024)
Lauraceae	Cryptocarya sp1 (S3036)
Lauraceae	Cryptocarya sp2 (S3185)
Lauraceae	Cryptocarya sp (S3202)
Lauraceae	Endiandra sp1 (S3094)
Lauraceae	Litsea sp2 (S3060)
Lauraceae	Neolitsea sp. (S3141)
Leeaceae	Leea indica (S3134)
Loganiaceae	Fagraea bodenii (S3055)
Loganiaceae	Neubugia rumphiana (B006)
Lomariopsidaceae	Bolbitis heteroclita (Presl.) Ching (B005)
Lomariopsidaceae	Lomagramma sinualata (B001)
Lomariopsidaceae	Lomaropsis kingii (Copel.) Holtum (B172)
Malvaceae	Hibiscus archboldiana (S3069)
Marantaceae	Donax cannaeformis (S3135)
Marantaceae	Maranta sp (S3001)
Marantaceae	Phrynium sp (S3134)
Melastomataceae	Medinella crassinervia (S3205)
Meliaceae	Aglaia cuculata (K3133)
Meliaceae	Aphanamixis polystachya (S3181)
Meliaceae	Aphanamixis sp (S3135)
Meliaceae	Chisocheton seramicus (B025)
Meliaceae	Chisocheton sp (S3022)
Meliaceae	Dysoxylum sp (S3190)
Meliaceae	Toona sureni (S3048)
Menispermaceae	Stephania japonica (S3072)
Menispermaceae	Tinomiscium petiolare (B372)
Menispermaceae	Tinospora dissitiflora (S3106)

Monimiaceae	Kibara sp (B067)
Monimiaceae	Kibara sp1 (B303)
Monimiaceae	Steganthera sp (B059)
Moraceae	Antiaris sp (S3121)
Moraceae	Artocarpus altilis (K3039)
Moraceae	Ficus nasuta (S3063)
Moraceae	Ficus nodiflora (S3247)
Moraceae	Ficus septica (K3063)
Moraceae	Ficus sp (S3149)
Moraceae	Ficus subulata (S3082)
Moraceae	Ficus virens (K3038)
Moraceae	Maclura amboinensis (B035)
Myristicaceae	Horsefieldia sp. (S3064)
Myristicaceae	Myristica acuminate (S3242)
Myristicaceae	Myristica insipida (S3077b)
Myristicaceae	Myristica sp (S3129)
Myrsinaceae	Ardisia sideromalla (S3102)
Myrsinaceae	Rapanea sp. (S3137)
Myrtaceae	Acmena accuminatissma (S2040)
Myrtaceae	Octamyrtus behrmannii (S3130)
Myrtaceae	Syzygium goniopterum (Diels)
(S3021)	
Myrtaceae	Syzygium sp. (S3252)
Myrtaceae	Syzygium sp1 (S3028)
Myrtaceae	Xanthomyrtus sp (S3210)
Nyctaginaceae	Pisonia sp (B036)
Ochidaceae	Gramphatophyllum sp (B181)
Oleaceae	Chionanthus sp (S3033)
Orchidaceae	Bulbophyllum sp (S3061)
Orchidaceae	Bulbophyllum sp1 (S3226)
Orchidaceae	Bulbophyllum sp2 (S3111)
Orchidaceae	Coelogyne sp (S3160)
Orchidaceae	Dendrobium sp (S3162)
Orchidaceae	Podochilus sp (S3017)
Pandanaceae	Freycinetia marantifolia (S3127)
Pandanaceae	Freycinetia sp (S3155)
Pandanaceae	Pandanus sp (S3016)
Piperaceae	Piper betle (S3150)
Piperaceae	Piper canicum (S3151)
Piperaceae	Piper ledermanii (S3229)
Piperaceae	Piper triangulare (S3231)
Piperaceae	Piper truncatibaccum (B095)
Pittosporaceae	Pittosporum sinuatum (K3150)
Poaceae	Bambusa sp (B565)
Poaceae	Ichnanthus vicinus (B852)
Poaceae	Leptaspis sp (S3227)
Polygalaceae	Xanthophyllum papuanum (S3099)
Polypodiaceae	Drymoglossum fallax (B362)
Polypodiaceae	Drymoglossum sp (B668)
Polypodiaceae	Goniophlebium sp (B157)
Polypodiaceae	Lecanopteris sirosa (B049)
Polypodiaceae	Microsorium cromwellii (B052)
Polypodiaceae	Microsorium glossophyllum (B002)
Polypodiaceae	Microsorium sp1 (B171)
Polypodiaceae	Polypodium drymnoglossum (S3208)
Polypodiaceae	Pyrrosia sp (S3003)
Proteaceae	Gevuina papuana (S3041)
Proteaceae	Helicia sp (S3100)
Proteaceae	Sternocarpus sp (S3248)
Ranunculaceae	Clematis clemensia (B018)
Rhamnaceae	Ziziphus angustifolia (K3125)
Rubiaceae	Anthocephalus chinensis (B849)
Rubiaceae	Canthium magistocaryum (K3101)
Rubiaceae	Ixora sp (S3236)
Rubiaceae	Lucinaea sp (S3236)
Rubiaceae	Musaenda pilosissima (S3209)
Rubiaceae	Myrmecodia sp (S3126)
Rubiaceae	Neonauclea sp (S3253)
Rubiaceae	Pachystylus gulcherianus (S3129)
Rubiaceae	Psychotria hollandiae (S3023)
Rubiaceae	Tarenna butensis (S3030)
Rubiaceae	Timonius timon (K3029)
Rubiaceae	Uncaria pendiculata (S3176)
Rubiaceae	Uncaria pendiculata (S3126)
Rubiaceae	Versteegia cauliflora (K3239)
Rutaceae	Evodia anisodora (S3122)
Rutaceae	Lunasia amara (S3071)
Sapindaceae	Harpullia pinnata (B159b)
Sapindaceae	Pometia pinnata (S3183)
Sapindaceae	Sarcopteris sp. (S3131)
Sapotaceae	Palaquim sp (B064)
Sapotaceae	Planchonella anteridifera (S3057)
Sapotaceae	Planchonella sp (S3255)
Selaginellaceae	Selaginella sp (S3141)
Smilacaceae	Smilax lancaefolia (S3019)
Smilacaceae	Smilax sp (K3083)
Sterculiaceae	Pterygota sp (S3250)
Sterculiaceae	Sterculia sp (S3110)
Symplocaceae	Symplocos sp. (S3180)
Tectariaceae	Tectaria beccariana (S3144)
Tectariaceae	Tectaria plesora (S3221)
Theaceae	Ternstroemia merrilliana (S3107)
Thelypteridaceae	Sphaerostephonos alatellus (S3232)
Thelypteridaceae	Thelypteris sp. (S3219)
Tiliaceae	Microcos sp (S3128)
Ulmaceae	Celtis latifolia (B084)
Ulmaceae	Gironniera sp (S3173)
Ulmaceae	Solenostigma sp (S3105)
Urticaceae	Elatostema sp (S3154)
Urticaceae	Elatostema sp (S3067)
Urticaceae	Fleurya sp (S3062)
Urticaceae	Poikilosperma sp (S3152)
Verbenaceae	Teijsmanniodendron sp (S3123)
Violaceae	Rinorea sp (S3092)
Vitaceae	Cissus sp (B389)
Winteraceae	Belliolum sp (S3230)
Zingiberaceae	Alpinia sp2 (S3137)
Zingiberaceae	Alpinia sp1 (S3137)
Zingiberacee	Amomum aculeatum (B099)

Proceedings of the Biological Society of New Guinea
R. Höft (ed.)

VEGETATION ANALYSIS OF THE WAENA - POLIMAK FORESTS; THEIR COMPOSITION, STRUCTURE AND ECONOMIC VALUE

Rosye H. R. Tanjung

University of Cenderawasih.

INTRODUCTION

Since primitive times humans have had the desire to understand the natural world. We know that humans as a part of the ecosystems, always have contact with their environment (plant, animal, etc). Thus, we continually make every effort to study characteristics of the natural environment.

Vegetation studies have been undertaken for a long time, originally due to an interest in the nature because vegetation is the basis of all life on earth.

Orlocí (1978) believed that in any vegetation study the objectives tend to be gathering informations on:

- the spatial arrangement of individuals or species on the ground and/or along the vertical profile of the community;
- trends and classes of variation in the similarity or dissimilarity in relation to species or entire stands;
- Population processes affecting spatial and/or temporal pattern in an area;
- Vegetation response to environmental influences and the dynamics of such a response.

In this preliminary work, I try to present quantitative data of the Polimak - Waena forests. Data was collected by using a combination of the transect and the nested plot methods.

The character of vegetation was described and analyzed following the methods described by Mueller-Dombois & Ellenberg (1974) and William (1986).

AIMS OF THE STUDY

Objective Aims

This preliminary study aims at presenting data relating to the species' presence, their distribution and economic value.

Subjective Aims

The subjective aims of this study are:

a. The lack of vegetation data of low land forests in the Jayapura area
b. The study site is an important area for conservation being situated on the border between the nature reserve of Cyclop and a settlement area.
c. The area under study plays an important role in the water source, erosion, and air pollution regulation for Jayapura and is habitat of some wild animals.

Terminology and definitions

Tropical low land forest: Tropical lowland forest is defined as tropical forests situated within the tropics at altitude below 1000 metres above sea level.

Primary and secondary forest: Primary and secondary forests are related to the condition of the community and the species composition showing the level of disturbance.

Tree and sapling: Trees and saplings are distinguished by their stem diameter. Their respective abundance refers to the regeneration of tree species and shows the structure of a community. In this study, trees are defined as woody plant (excluding shrubs) with a diameter exceeding 10 cm and sapling refers to tree species with diamter between 2 cm and 10cm.

STUDY AREA

Location

The study area occupies an area of lowland forest at Polimak - Waena in Jayapura, approximately 10 km south of the town of Jayapura.

Study sites are located at approximately 300 to 400 metres above sea level. This part of Jayapura is considered to be the border line between the settlement area of Polimak - Waena and the nature reserve of Cyclop.

Vegetation

The vegetation surrounding the study area is dominated by lowland vegetation ranging from grassland, shrubland and secondary forest to primary forest. The distribution of the these types of vegetation depends on the location and the distance from the settlement area. Most of the area close to the settlements are dominated by grasses, shrubs and cultivated land.

Topography

The study site consists of a gently sloping terrain with small hills interspaced with level plains. The Polimak and Entrop Rivers cross the study sites. Kamp Walker River meanders along the southern and eastern side with the Cyclop Mountains lying to the west and the Polimak - Waena settlement being situated in the east.

MATERIAL AND METHODS

The study commenced in March 1993. The sampling technigue adopted was the transect method. The study sites were selected following a gradient according to the procedure outlined by Tansley (1946); Mueller-Dombois & Ellenberg (1974) and Moore & Chapman (1986). The study was carried out in the lowland rain forest of Polimak - Waena, Jayapura.

Two types of sample plots were used. Circles with areas of 16 m² were employed to collect sapling data and 100 m² circles were used in the vegetation survey of trees (Oosting 1956; Lockett unpublished).

Quantitative measurements of the vegetation present in each plot were made, including the diameter of stems and their density. The diameter of trees was measured at 4.5 feet above ground level (Myer & Shelton 1980; Avery & Burkhat 1983).

Analysis of Data

Quantitative data was calculated following the formula below (after Mueller-Dombois & Ellenberg 1974):

$$\text{Relative Density} = \frac{\text{Density of each species}}{\text{Total Density of species}} \times 100\ \%$$

$$\text{Relative Frequency} = \frac{\text{Frequency of each species}}{\text{Total frequency of species}} \times 100\ \%$$

$$\text{Dominance} = \frac{\text{Basal area of each species}}{\text{Total basal area}}$$

$$\text{Relative Dominance} = \frac{\text{Dominance of each species}}{\text{Total Dominance of species}} \times 100\ \%$$

$$\text{Importance Value} = \text{RDe} + \text{RF} + \text{RDo}$$

Relative Density (RDe), Relative Frequency (RF) and Relative Dominance (RDo) were calculated for each species at every site. These values were used to calculate the importance value (IV). The maximum importance value is 300 percent and it only occurs in uniform vegetation (dominated by one species). The Relative Dominance was calculated from the basal area.

The relative frequency value was used to consider the distribution of species of the area in the quadrat samples. Species with high importance values may be used to predict characteristics of the composition of the vegetation in future and the effects of disturbance (Mueller-Dombois & Ellenberg 1974).

An Index of Similarity (IS) was calculated for each site comparing it to the other sites. If the IS was less than 50 percent, it could be assumed that the two sites differed dramatically in their composition.

The community index was based on the methods of Sorensen cited in Mueller-Dombois & Ellenberg (1974). It is based on the presence or absence of species. The formula applied was:

$$cc = \frac{2\,c}{a + b} \times 100\ \%$$

where a = Total number of species on the Waena site, b = Total number of species on the Polimak site, c = Number of species common to the two sites, and cc = the index of similarty.

RESULTS AND DISCUSSION

Waena Site

Thirty one tree species were recorded at the Waena site. Three of the main species were Burseraceae. This family seemed to be well represented in the area. *Haplolobus floribundus* and *Canarium oleosum* had a high density (8.9 %) follwed by *Aphania*, *Pometia pinnata* and *Chisocheton* sp. which have same relative density (6.7 %). The other main species had relative densities lower than 5 percent.

According to the frequency of appearance in the plot sample *Haplolobus floribundus*, *Aphania senegalensis* and *Pometia pinnata* had the highest relative frequency (7.14 %) follwed by *Canarium oleousum*, *Chisocheton*, *Inocarpus papuanus*, *Canarium asperum*, *Eugenia* sp. and *Paratocarpus* sp. (4.76 %).

Due to variations of the diameter, species with high density or high frequency are not necessarily the most important species. Table 1 shows the effect of density, frequency and diameter (dbh) on the importance value (IV) for the most important species in the area.

Table 1. Important tree species from the Waena site with their relative densities (RDe) relative frequencies (RF), relative dominance (RDo) and importance values (IV)

No.	Species	RDe	RF	RDo	IV
1.	*Haplolobus floribundus*	8.90	7.14	17.72	33.40
2.	*Aphania senegalensis*	6.70	7.14	12.61	26.45
3.	*Canarium oleosum*	8.90	4.76	4.20	17.50
4.	*Chisocheton* sp.	6.70	4.76	6.00	17.46
5.	*Pometia pinnata*	6.70	7.14	3.30	17.14
6.	*Inocorpus papuanus*	4.23	4.76	6.00	14.99
7.	*Canarium asperum*	4.23	4.76	5.40	14.39
8.	*Eugenis* sp. A	4.23	4.76	3.90	12.89
9.	*Eugenia* sp. B	4.23	4.76	3.03	12.29
10.	*Paratocarpus* sp.	4.23	4.76	2.42	11.41

Only one of the main species, *Pometia pinnata*, is edible. *Pometia pinnata* commonly bears fruits between December and February. Collecting these fruits is an important reason for people to get into the forest and effect the vegetation. Other species produce wood of low to medium timber quality.

According to the sapling records, regeration of the vegetation is good. 59 species of saplings were recorded on the Waena site. *Haplolobus floribundus* and *Pometia pinnata* were found to reproduce readily with the saplings having high relative density, relative frequency, relative dominance and consequently high importance values. (Table 2). The ooverall density of samplings was relatively low (690/ha) while that of tree was 336/ha.

Table 2. Important species of saplings from the Waena site with their relative densities (RDe), relative frequencies (RF), relative dominance (RDo) and importance values (IV).

No.	Species	RDe	RF	RDo	IV
1.	*Haplolobus floribundus*	48.93	12.03	44.4	117.36
2.	*Pometia* sp.	10.10	12.90	14.14	37.24
3.	*Eugenia* sp. A	11.22	9.67	12.12	32.01
4.	*Terminalia melanocarpa*	6.12	9.67	8.08	24.87
5.	*Mischocarpus* sp.	6.12	9.67	6.06	22.85
6.	*Canarium asperum*	5.10	6.45	3.03	14.58
7.	*Canarium oleosum*	3.06	6.45	4.04	13.55
8.	*Macaranga* sp.	3.06	6.45	3.03	12.54
9.	*Eugenia desinflora*	3.06	6.45	2.02	11.53
10.	*Inocarpus papuanus*	3.06	3.22	3.03	9.31

Table 2 shows that *Haplolobus floribundus* (Burseraceae) regenerated well at the Waena site. *Haplolobus floribundus* tended to be the dominant species in the sapling stage as well as in the tree layer (Table 1).

In the sapling stage *Macaranga* sp., which was a common species in disturbed areas, had a relatively high importance value. It was found in areas with an open canopy. This showed that part of the area had been disturbed, naturally or as an effect of human activities.

Both, the tree and sapling densities as well as the species diversity were low compared to other areas as recorded by Steenis (1957), Whitmore (1975) and Anggana (1977).

The diameters of tree stems were relatively small. Most trees had diameters between 10 and 20 cm. It seems that the area is still developing to reach the mature forest stage. However, the species compositon has revealed common features of a mature non-dipterocarp forest at an altitude below 1000 m.

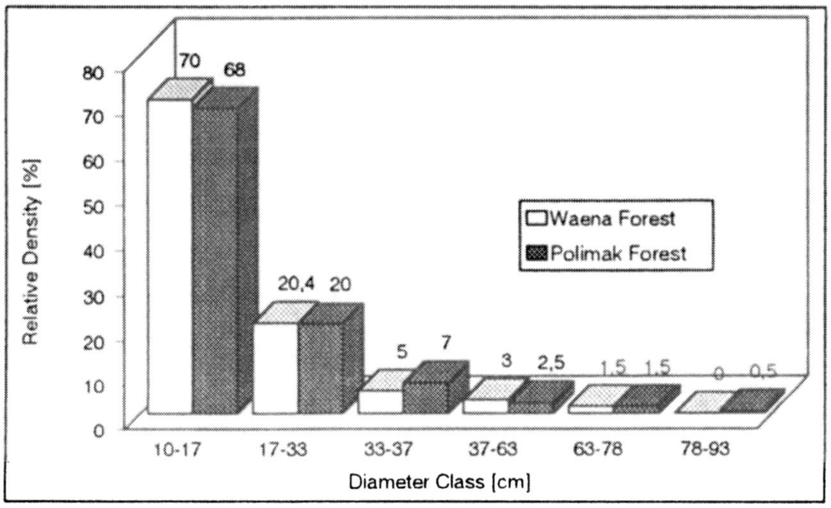

Figure 1. Diameter class distribution of trees from the Waena and Polimak sites (Harston 1976)

The smallest diameter class had the highest density (70 %) followed by second class (20.4 %). Individual numbers decreased dramatically within the larger diameter classes. No tree with a stem diameter above 78 cm was found.

Polimak Site

Table 3. Important tree species from the Polimak site with their relative densities (RDe), relative frequencies (RF), relative dominance (RDo) and importance values (IV)

No.	Species	RDe	RF	RDo	IV
1.	*Canarium oleosum*	15.52	11.8	10.01	37.33
2.	*Haplolobus floribundus*	12.97	8.9	12.53	34.40
3.	*Eugenia densiflora*	5.1	5.9	6.03	17.03
4.	*Pometia* sp.	5.1	5.9	6.03	17.03
5.	*Sloanea elongata*	5.1	5.9	5.95	16.95
6.	*Paratocarpus* sp.	5.1	3.9	4.65	15.65
7.	*Alstonia scholaris*	5.1	5.9	4.32	15.32
8.	*Imelio dendron*	5.1	5.9	4.22	15.22
9.	*Aglaia cuspidata*	5.1	5.9	4.09	15.09
10.	*Mischocarpus* sp.	5.1	5.9	3.29	15.09

On the Polimak site, only 27 tree species and 64 species of saplings were recorded. In some plots the number of tree species decreases due to the association with a palm and *Pandanus* sp.

Table 3 presents the important species of the Polimak forest.

Similarly to the Waena site, Burseraceae tend to be the most important species. The species composition is variable, however, the set of species found on the two sites is relatively similar (IS=65.5%).

Figure 1 shows that most trees have stem diameters ranging from 10 to 17 cm. Similar to the Waena site, the density and the diversity of species is low (337 trees per hectare and 982 saplings per hectare).

Table 4. Important species of saplings from the Polimak site with their relative densities (RDe), relative frequencies (RF), relative dominance (RDo) and importance values (IV).

No.	Species	RDe	RF	RDo	IV
1.	*Eugenia densiflora*	8.72	6.38	11.45	26.55
2.	*Chisocheton* sp.	6.71	4.25	6.48	17.44
3.	*Paratocarpus* sp.	4.03	3.19	6.48	13.70
4.	*Haplolobus floribundus*	4.70	2.12	6.26	13.08
5.	*Canarium oleosum*	4.03	4.25	3.37	11.95
6.	*Ficus* sp.	4.03	4.25	2.81	11.09
7.	*Eugenia* sp. C	3.36	1.06	4.32	8.74
8.	*Inocarpus papuanus*	2.68	3.19	2.38	8.25
9.	*Pometia pinnata*	2.68	3.19	2.16	8.03
10.	*Canarium* sp. A	2.01	3.19	2.38	7.58

Table 4 shows that the saplings of *Eugenia* sp. have the highest importance value. There is little difference in species composition to the Waena site.

Regeneration among certain species including *Eugenia densiflora*, *Haplolobus floribundus* and *Paratocarpus* sp. is well on this site.

CONCLUSION

The data presented in this study indicates that both species composition and structure of the vegetation of Polimak and Waena forests is relatively similar. The sites are classified as immature non-dipterocarp lowland forest.

Species of Burseraceae, Myrtaceae, Sapindaceae and Meliaceae dominate among the trees and saplings. *Macaranga* which is pioneer species was found to regenerate well in open areas. It is an important species of the secondary forest surrounding the study area.

Pometia pinnata was found to be well distributed throughout the area. It has been noted before that *Pometia pinnata* produces fruits consumed by people in Irian Jaya. This is considered to be an important reason for people to get into the forest and consequently effect the vegetation to change.

The reason for the low density and low diversity of species is unknown. There is an obvious need for more detailed work on the soil conditions, the soil seed bank and seed viability, as well as on the characteristics of some species and the levels and types of disturbance.

Suggestions for further studies

Some work could be done by marking out and observing permanent plots in terms of:

- Regeneration patterns of some interesting and important species;
- Floristic changes;
- Litter production during the wet and the dry season..

REFERENCES

Armesto, J.J. & Fuentes, E.R. (1988). Tree species regeneration in a mid-elevation temperate rain forest in Isla de Chiloe, Chile. *Vegetatio* **74,** 151-159.

Auld, J.D. (1980). Population dynamics of the shrub *Acacia suaveolens* (Sm.) Willd. *Australian Journal of Ecology* **2**, 373-385.

Avery, T.E. & Burkhat, H.E. (1983). *Forest measurement.* Third edition, McGraw-Hill Book Company. Toronto.

Connell, J.H. & Slatyer, R.O. (1977). Mechanism of succession in natural communities and their role in community stability. *American Naturalist* **111,** 1119- 144.

Engler, F.E. (1954). The study of plant communities. *An introduction to plant ecology* (Ed. by W.H. Freeman), San Fransisco, London.

Hara, M. (1987). Analysis of seedling banks of climax beech forest: ecological importance of seedling sprouts. *Vegetatio* **71,** 67-74.

Hartshorn, G. Tree falls and tropical forest dynamics. *Tropical trees and living systems.* (Ed. by P.B. Tomlinson & M.H. Zemmirans),Cambridge University Press. London.

Lockett, E.J. (unpublished). The interpretation of regeneration survey data. Tasmanian Forest Commission. Australia.

Mueller-Dombois, D. & Ellenberg, H. (1974). *Aims and methods of vegetation ecology.* Wiley. New York.

Moore, P. & Chapman, S.B. (1986). *Methods in plant ecology.* 2nd edition. Blackwell Scientific Publishers. Melbourne.

Myer, W.L. & Shelton, R.L. (1980). *Survey methods for ecosystem management.* A Wiley Interscience Publication, John Wiley & Sons, Brisbane.

Niering, W.A. & Engler, F.E. (1955). A shrub community of *Viburnum lentago* stable for twenty-five years. *Ecology* **36,** 356-360.

Niering, W.A. & Goodwin, R.H. (1974). Creation of relitively stable shrublands on righ of way and pasture land. *Ecology* **55,** 784-795.

Orloci, L. (1977). An information theory model for pattern analysis. *J. Ecology* **59,** 343-349.

Odum, E.P. (1986). The strategy of ecosystem development. *Science* New York **164,** 262-270.

Oosting, H.J. (1956). The study of plant communities. *An introduction to plant ecology.* (Ed. by W.H. Freeman), San Fransisco, London.

Sastrapraja, S. (1980). *Kayu Indonesia.* Balai Pustaka, Jakarta.

Saryono, H. (1980). *Ekonomi Kehutanan Indoneisa.* Yasaguna, Jakarta.

Steenis.C.C.G.J. (1957). An outline of the vegetation types in Indonesia and some adjacent regions. *Proc. 8. Pacific Science Congress* **4,** 61-97.

Tansley, A.G. (1946). *Introduction to plant ecology.* George Allen & Unwin Ltd. London.

William, L.D. (1986). *The Ecology of the Forest and Woodland of South Australia.* (Ed. by H.R. Wallace & D.J. Woolman), Government Printer. South Australia.

Whitmore, T.C. (1975). *Tropical rain forests of the Far East.* Claredon Press. Oxford.

APPENDIX 1

List of species (diameter ≥ 10 cm) found at the Polimak site

Aglaia cuspidata	Eugenia densiflora	Michocarpus
Albizia sp.	Eugenia sp. A	Pometia pinnata
Alstonia scholaris	Eugenia sp. B	Nauclea sp.
Aphania senegalensis	Ficus nodusa	Pometia sp.
Canarium oleosum	Ficus sp. A	Paratocarpus sp.
Canarium asperum	Haplolobus floribundus	Sloetia elongata
Chisocheton schumanii	Imeliodendron sp.	Talauma sp.
Dillenia sp.	Intsia sp.	Terminalia sp.
Dysoxylum arboresum	Macaranga fimbriata	Xylocarpus sp.

APPENDIX 2

List of tree species (Diameter ≥ 10 cm) found at the Weana site

Aglaia cuspidata	Dysoxylum aboressuns	Macaranga fimbriata
Alstonia scholaris	Dysoxylum sp.	Mischocarpus sp .
Aphania senegalensis	Eugenia sp. A	Myristica sp.
Canarium olesum	Eugenia densiflora	Nauclea sp.
Canarium asperum	Eugenia sp. B	Paratocarpus sp .
Canarium sp. A	Eugenia sp. C	Pometia pinnata
Canarium sp. B	Ficus nodusa	Pometia sp.
Canarium sp. C	Haplolobus floribulus	Sloetia elongata
Chisocheton schumannii	Imeliodendron sp.	Trichodenia philipensis
Callophyllum tuberrosum	Inocarpus papuanus	
Cryptocarya crassinervia	Intsia sp.	

Proceedings of the Biological Society of New Guinea
R. Höft (ed.)

CHARACTERISTIC EPIPHYTE SPECIES OF THE MONTANE FORESTS IN MOROBE PROVINCE, PAPUA NEW GUINEA

Robert Höft & Martina Höft [†]

Wau Ecology Institute, P.O. Box 77, Wau, Papua New Guinea
[†] *University of Bayreuth, Postfach 101251, 95400 Bayreuth, Germany*

INTRODUCTION

Epiphytes have attracted the attention of biogeographers (Schimper 1888; Went 1940 etc.) and ecologists (e.g. Benzing 1983) since the early times of botanical exploration of the tropics. However, these studies were usually seen as being of purely academic interest. Until today, epiphytes are considered not to have any significant economic value (except for a small market for ornamental plants) and their role in the ecosystem is often neglected.

The range of available herbarium material of epiphytic plants traditionally suffers from the special efforts necessary for their collection. It is therefore not astounding that epiphytic plants are among the least studied vascular plants taxonomically despite the fact that a quarter of the higher plant species occuring in New Guinea are - at least facultatively - epiphytic.

Epiphytes have received some attention from ecophysiologists (one such study was carried out in Papua New Guinea by Earnshaw *et al.* 1987) but little work has been done on their role within the ecosystem (see for example Nadkarni & Matelson 1991).

Descriptions of epiphytes on community level are even more scarce. An analysis of epiphyte communities was first attempted only recently in Central American montane rain forests (Catling & Lefkovitch 1989). Similar studies have been performed for cryptogamic epiphyte communities (Cornelissen & Ter Steege 1989; Kürschner 1990).

Our study attempts at delineating epiphyte communities based on a purely floristic analysis and at describing "character epiphytes", *i.e.* species that typically occur in a certain forest type and not or rarely in others.

METHODS

Data was collected on the slopes of Mt. Kaindi and along the field station of Wau Ecology Institute in the Kuper Range near Mt. Kolorong. A total of 20 trees from various vegetation types and elevations between 1100 and 2350 m a.s.l. were selected subjectively according to their epiphyte load. The single-rope technique (Perry 1978) provided access to all parts of the canopy. Vascular plants on the trunks and branches of each tree were recorded and their cover/abundance was estimated using the scale introduced by Braun-Blanquet as described in Mueller-Dombois & Ellenberg (1974). Unknown species were collected and identified as far as possible at the National Herbarium in Lae. Working names were assigned to the species reoccuring but not yet identified. The total cover of mosses and lichens was recorded. Cryptogams, however, were neither collected nor identified.

A computerised program for "hand-sorting" the relevés (Durka & Ackermann 1992) was used for the initial depiction of the data set. As soon as the safe identification of the mostly sterile material has progressed (by revisiting the sites at various times of the year and recollection of the then fertile material) the actual numerical analysis will be carried out.

Classification and ordination methods [especially Principal Component Analysis (PCA) and Canonical Correlation Analysis (Canoco) with the elevation as key parameter] will be used.

Preliminary Results

According to the present analysis, 193 species of epiphytes were collected from the canopies and 105 species from the trunks of the 20 trees under investigation. To date, some 60 percent of the species could be identified to species.

Floristically, three distinct epiphyte communities could be delineated (Tab. 1)

Table 1. Constancy table of the character species of submontane, mid-montane and upper mid-montane forest with categories of frequency (first figure; V = occurring in 80 to 100 percent of the relevés, IV = in 60 to 80 percent, III = in 40 to 60 percent, II = in 20 to 40 percent, I = in less than 20 percent; arab instead of roman figures are used when summarising less than 5 relevés) **and mean abundance** (second figure; R = rare, + = frequent but little cover, 1 = covering 1 to 5 percent of the relevé area, 2 = covering 5 to 25 percent, 3 = covering 25 to 50 percent, 4 = covering 50 to 75 percent, 5 = covering 75 to 100 percent). Numbers of unidentified species refer to the authors' collection presently held at Wau Ecology Institute

Forest Type	**Submontane**	**Mid-Montane**	**Upper Mid-Montane**
Number of Relevés per Community	10	6	4
Average Number of Species per Relevé	18	34	25
Mean Elevation	1449	2057	2235
Character Species of Sub-Montane Forests			
Belvisia cf. *mucronata* 2568	III 1		.
Liparis cordylobulbon Rchb. f.	II +	.	.
Peperomia tetraphylla (Forst. f.) Hook	II +		.
Piper sp. 2587	I 2	.	.
Eria imitans Schltr.	I 1	.	.
Platycerium wandae Racib.	I 1	.	.
Pyrrosia cf. *varia* 2752	I 1	.	.
Dendrobium pleianthum Schltr.	I 1	.	.
Asplenium nidus L.	I +	.	.
Coelogyne fragrans Schltr.	I +	.	.
Oberonia sp. A 2714	I +	.	.
Belvisia mucronata (Fée) Copel.	I +	.	.
Character Species of Mid-Montane Forests			
Vaccinium horizontale Sleum.		V 1	.
Glossorhyncha aff. *pendulata* van Royen	.	V +	.
Rhododendron gracilentum F. Muell.	.	V +	.
Dendrobium terrestre J.J. Sm.		V 1	.
Glomera platypetala Schltr.		IV 1	.
Humata deltoidea Copel.		IV +	1+
Elaphoglossum cf. *sclerophyllum* v.A.v.R.	.	IV +	
Dimorphanthera cf. *leucostoma* 3461	.	IV +	.
Rhododendron lindaueanum Koord. var. *lindaueanum*.		IV +	.
Dendromyza reinwardtiana (Bl. ex Koord.)		III 1	.
Bulbophyllum sp. A 3333		III +	.
Vaccinium sp. A 3339		III +	.
Ctenopteris denticulata (Bl.) C. Chr		III +	.
Dendrobium cf. *subclausum* Rolfe		III +	.
Vaccinium sp. B 3354		III +	.
Selliguea aff. *brachylepidota* 2807	.	II 1	.
Agrostophyllum cf. *graminifolium* 2835	.	II +	.
Bulbophyllum sp. C 2831		II +	.
Dendrobium cf. *kenejianum* Schltr.	.	II +	.
Meringium cernuum Gepp	.	II +	.

Diplycosia morobeensis Sleum.	.	II +	.
Glossorhyncha cf. *scopulata* van Royen	.	II +	.
Ctenopteris cf. *micropaleata* 3765	.	II +	.
Glomera nigrilimbata van Royen	.	II R	.
Agrostophyllum cf. *patentissimum* 3504	.	II R	.
Vittaria elongata Sw.	.	II R	.
Character Species of Mid-Montane and Upper Mid-Montane Forests			
Selliguea brachylepidota Copel.	.	IV 1	3 1
Humata deltoidea Copel.	.	I R	3 +
Dendrobium aff. *dulce* J.J. Sm.	.	I 3	2 1
Phreatia valida Schltr.	.	I R	1 +
Character Species of Upper Mid-Montane Forests			
Glossorhyncha sp. A 2885	.	.	2 1
Mediocalcar sp. C 2896	.	.	2 +
Dendrobium cf. *neuroglossum* Schltr.	.	.	2 R

Some 15 percent of the almost 300 species of epiphytes were restricted to a certain altitudinal range and thus characterised a specific forest type. The largest number of character species (27) was found in the mid-montane forest. This forest type was also the most species rich in terms of total numbers of epiphytic species (34 epiphyte species per tree in average).

The sub-montane forest (11 character species; 18 epiphyte species per tree in average) and the upper mid-montane forest (3 character species; 25 epiphyte species in average) were less clearly defined. Four species occurred both in the mid-montane and the upper mid-montane forest thus seperating montane from submontane forests.

Among the character species, the percentage of orchids increased constantly with higher elevation. In the submontane forests, orchids made up 35 percent of the character species; 44 percent of the character species in the mid-montane forests were orchids and all three of the upper mid-montane character epiphytes belonged to the family Orchidaceae.

Ericaceae occurred mainly in the mid-montane forests and made up almost 30 percent of the character species in this forest type.

Overall, more than three quarters of the epiphytic species characteristic of a specific forest type belonged to the Pteridophyta and Orchidaceae.

Discussion

According to our data set, mid-montane forests in their strict sense (between 1900 and 2100 m a.s.l. in the study area) appear to be richer in epiphytic species than the forests of lower and those of higher elevations. Data on forest structure suggest, however, that the trees investigated in submontane forests and in the upper mid-montane forests partly originated from disturbed forest types, whereas the mid-montane forests appeared to be undisturbed, mature forest. A comparison of epiphyte species richness among the three forest types can thus not be made from our data since regrowth forest has been found to be gernerally poorer in epiphytic species than older forest (Whitmore 1974).

Other authors have found a correlation between precipitation (including fog) and epiphyte species richness (e.g. Wee 1978; reviews in Richards 1952 and Whitmore 1984) and non-vascular epiphytes have their highest abundance in the upper montane forest zone. Exact rainfall data are not available for the various collecting sites within the study area. Nevertheless, observations indicate that rainfall is higher in the mid-montane forest area (3100 mm at 2360 m elevation; cited after Gressit & Nadkarni 1978) than in the submontane forests (1864 mm at 1200 m elevation; see climate diagram in Höft *et al.*, this volume). The higher diversity of epiphyte species in the mid-montane forests is in consistence with the climatic data.

It is yet unclear, however, at which elevation the highest diversity of vascular epiphytes can be found in the study area. More details on the status of the forest are required to make sure that the respective vegetation types have undergone the same degree of disturbance and are thus comparable.

Further problems of data comparison arise from the subjective selection of trees. Observations indicate that the epiphyte species composition on dead phytophores (which have not been analysed for security reasons) may be different to the one on healthy trees and that the germination of some epiphytes can be facilitated on trees which already bear a load of epiphytes.

For these reasons, detailed biodiversity studies which include the epiphytic flora will require a large sample number and random selection of host plants if they attempt to be more than just incomplete species lists.

ACKNOWLEDGEMENTS

We thank Yves Basset for allowing us the use of his climbing gear. Martin Hutten contributed substantially to the completion of field work. Paul Katik and Neville Howcroft were invaluable in identifying some of the plant material. The study was partially funded by the A.F.W. Schimper scholarship of the H. & E. Walter foundation. Wau Ecology Institute supported all aspects of the study.

REFERENCES

Benzing, D.H. (1983). Vascular epiphytes: a survey with special reference to their interactions with other organisms. *Tropical rain forest; ecology and management.* (Ed. by S.T. Sutton, T.C. Whitmore, A.C. Chadwick), pp. 11-24, British Ecological Society, Special Publication No. 2.

Catling, P.M. & Lefkovitch, L.P. (1989). Associations of vascular epiphytes in a Guatemalan cloud forest. *Biotropica* **21(1),** 35-40.

Cornelissen, J.H.C. & Ter Steege, H. (1989). Distribution and ecology of epiphytic bryophytes and lichens in dry evergreen forest of Guayana. *J. Trop. Ecol.* **5(2),** 131-150.

Durka, W. & Ackermann, W. (1992). *Sort 2.1. A program for sorting vegetation tables.* University of Bayreuth.

Earnshaw, M.J., Winter, K., Ziegler, H., Stichler, W., Cruttwell, N.E.G., Kerenga, K., Cribb, P.J., Wood, J., Croft, J.R., Carver, K.A. & Gunn, T.C. (1987). Altitudinal changes in the incidence of crassulaceaen acid metabolism in vascular epiphytes and related life forms in Papua New Guinea. *Oecologia* **73,** 566-572.

Gressitt, J.L. & Nadkarni, N. (1978). *Guide to Mt. Kaindi. Background to montane New Guinea ecology.* Wau Ecology Handbook No. 5, 135 pp.

Höft, R., Lange, S., Spang, O. (this volume). The regeneration potential of a submontane grassland site in Wau, Morobe Province, Papua New Guinea.

Kürschner, H. (1990). Die epiphytischen Moosgesellschaften am Mt. Kinabalu (Nord-Borneo, Sabah, Malaysia). *Nova Hedwigia* **51(1-2),** 1-75.

Madison, M. (1977). Vascular epiphytes: their systematic occurrence and salient features.- *Selbyana* **2,** 1-13.

Mueller-Dombois, D. & Ellenberg, H. (1974). *Aims and methods of vegetation ecology.* New York, Wiley, 207 pp.

Nadkarni, N.M. & Matelson, T.J. (1991). Fine litter dynamics within the tree canopy of a tropical cloud forest. *Ecology* **72(6),** 2071-2082.

Perry, D.R. (1978). A method of acces into the crowns of emergent and canopy trees.- *Biotropica* **10,** 155-157.

Richards, P.W. (1952). *The tropical rain forest.* Cambridge University Press.

Schimper, A.F.W. (1888). Die epiphytische Vegetation Amerikas. *Bott. Mittheilungen aus den Tropen,* Jena, **2,** 1-162.

Wee, Y.C. (1978). Vasscular epiphytes of Singapore's wayside trees. *Gdns. Bull. Singapore* **31,** 114-126.

Went, F.W. (1940). Soziologie der Epiphyten eines tropischen Urwaldes. *Ann. Jard. Bot., Buitenzorg* **50,** 1-98.

Whitmore, T.C. (1974). *Change with time and the role of cyclones in tropical rain forest on Kolombangara, Solomon Islands.* Commonwealth Forestry Institute Paper 46.

Whitmore, T.C. (1984). *Tropical rain forests of the Far East.* Oxford University Press, Oxford, New York, Toronto.

Proceedings of the Biological Society 1993
R. Höft (ed.)

THE HIGHLAND VEGETATION OF BALIEM VALLEY, IRIAN JAYA

Rosye H.R. Tanjung

University of Cenderawasih, P.O. Box 422, Jayapura 99351, Indonesia

Ericaceae (*Rhododendron* and *Vaccinium*) and Fagaceae (*Nothafagus*) were found to be dominant between 2000 and 3000 metres above sea level. At about 3500 m a.s.l. the vegetation occured as mossy forest vegetation dominated by *Phyllocladus* and *Papuacedrus*. Several species of ground orchids associated with ferns as ground vegetation were found in the mountain forest type. On wet open areas occurring between 2500 and 3500 metres a.s.l. *Cyathea* became dominant.

The Baliem Valley forest will be a focus of future studies because of its diversity. Much work is expected to be done in various aspects of botany.

Proceedings of the Biological Society of New Guinea
R. Höft (ed.)

PHOTOCOPIES OF THE PLANTS OF THE PAPUA NEW GUINEA UNIVERSITY OF TECHNOLOGY CAMPUS

Jonathan C. Reid

Department of Agriculture, Papua New Guinea University of Technology, Private Mail Bag, Lae, Morobe Province, Papua New Guinea

ABSTRACT

Listed are 113 families to which are assigned 769 species collected so far from the Papua New Guinea University of Technology campus, Lae. Cultivated species (517) are much more numerous than wild (252) species, although this partly results because virtually all trees are classed as cultivated. Grass species (71) are numerically important among the herbs, with forb species (308) and tree species (209) being of primary importance over epiphytes (63), shrubs (62), and climbers (48). Families containing cultivated species number 106 as opposed to 54 containing wild species. Trends in numbers for families generally reflect those for species. Differences in numbers of species and families between cultivated and wild categories and between life forms, and, for species, between higher taxa are further considered.

A description is given of how photocopies were made of plant specimens to provide a convenient record supplementing herbarium material.

INTRODUCTION

The Papua New Guinea University of Technology (UNITECH) campus at Lae shares common vegetative features with campuses and their staff housing areas observed in Nigeria, particularly the University of Calabar in the southeast, and probably with those in much of the tropics.

At such sites there is much human disturbance of the vegetation. Indigenous species of the moist tropical lowland forest are to a large extent replaced by vagile, cosmotropical species, particularly among ruderal and cultivated (ornamental and crop) forms (for example, see Holm *et al.* 1979).

Experience and familiarity with the taxa in Nigeria was fairly easily transferred and applied to the flora of the UNITECH campus. This has allowed compilation of a list of the species at UNITECH; particularly because of the much better availability of resources for obtaining determinations at Lae.

OBJECTIVES

The study started with an attempt to make a rapidly accessed data base on plants as alternative hosts of economic insects; and to provide a ready guide to the weeds of the campus, as a supplement to the excellent guide Henty and Pritchard (1982) give for Papua New Guinea.

A consideration also was that it would be a contribution to the important requirement for updating botanical records. Anthropogenic vegetation has not generally been the major target for collectors here and recent collecting has not been intensive. With much of the campus population traveling out of the country every 18 months (about 200 staff members are expatriates), it seemed probable that exotic species to add to Papua New Guinea's list could be found; for example, two species of *Kalanchoe* (Crassulaceae) not listed by Höft (1992, p. 99). This may have implications for quarantine.

These aims became absorbed as the work suggested several other purposes, two of which are the subject of this paper.

1. The compilation, as is usual, depends on specimens on herbarium sheets, but at the same time it appeared possible to produce a comprehensive collection of photocopies of the specimens, a good deal handier for reference than herbarium sheets.
   This collection should enable easy identification of most of the plants in this and similar floras. It is nearly complete but needs notes and identifications written on the sheets. Hopefully it can be published in a form which allows additions.
   To this end the major techniques for photocopying learnt during the exercise are included in the methods here.
2. The list allows a taxonomic analysis of the community. Analyses of campus or other quasi-urban floras are scarce in the literature, despite their commonness and the obvious availability of such floras for use by academics.

The numbers of species are tabulated for families and higher taxa for cultivated and wild vegetation within the life forms: herbs epiphytes, climbers, shrubs and trees. These will subsequently be analyzed ecologically and taxonomically and compared with less artificial floras to learn how they differ.

That the list (as of 9.11.1993) is as yet incomplete is possibly an objection to publishing it now. However, the number of species collected per unit effort is approaching an asymptote. In view of the frequency with which an assessment of a flora must by necessity be made from a superficial survey, as in urgent conservation programmes, it may be useful to record the list at this stage. Later can be learnt whether there was very much value in obtaining the extra information in the eventually exhaustive list when they are compared.

MATERIALS AND METHODS

Site

UNITECH campus is 8 km from the city of Lae in the Morobe Province, at mostly 50-70 m a.s.l. The land is flat and lies about 1 km east of the Bumbo River. Its extent is about 2.5 km^2 (about 2.7 km long and 1.4 km at its widest).

Gurnah (1985) extracted these data for the site. "The soil is a Troporthent (Bleeker 1983), light sandy clay loam overlying deep gravel and stones... Mean annual rainfall is 4112 mm, with the driest month getting no less than 200 mm... Mean relative humidities at 09.00 h and 15.00 h are 82% and 71% respectively".

There are effectively no natural or large permanent water bodies on campus except the sewage ponds, although a characteristic ruderal hydrosere develops in and near wheel ruts.

Reid (in prep.) estimates the population of the campus at around 3000. He mentions aspects of housing in considering the vegetation removal regime.

A few specimens collected from off the campus or their photocopies were better than others obtained from campus representatives of the species. Thus, their photocopies may be used in preference. Several of those off campus specimens, along with some phena now lumped with others for the present, were also counted for the oral presentation of the list. That should be replaced by this present one which excludes them.

Collecting

Collecting essentially began in January, 1993 and has continued with about the same overall intensity until today. An average of about an hour a day is expended on collecting and the photocopies. Herbarium specimens have probably taken up as much time and preliminary identifications about the same. The number of species collected (including those from off campus which do not affect the trend) is approaching an asymptote. Thus probably all but some 5-10 percent of the species and presumably virtually all of the families have been collected. Species distinguished but so far sterile or with inaccessible flowers and fruit probably comprise the bulk of that percentage. Some species were transplanted to my garden to provide flowers or fruit (preferably both) and a score or so of specimens from collections off campus were planted and still survive in the garden. These are included in the count.

Five or more observations have been made of most gardens on the campus. Most of the common grounds have been visited more frequently. No refusal of a request to look for or take a specimen was received in hundreds of gatherings.

Specimens were preserved generally according to usual herbarium techniques but particularly were photocopied. Sufficient material for five or six duplicates or often less was gathered (folded if necessary). The part of the material destined for photocopying was made to fit approximately A4 (210 x 297 mm) size, rather than larger herbarium sheet size. It was arranged as well as possible into a horizontally folded newspaper sheet (to be essentially A4 size) inside a file folder. In the field it was convenient to have up to 20 or so such files available and carry them loose but under sufficient pressure to flatten, but not greatly crush and crease, the specimens.

Photocopying

Photocopies were usually made before the material was more than one day (occasionally up to a week) old, although there is often little difference between copies of dry and fresh material.

Copies were made on A4 standard photocopy paper (from SFI, 80 gms), virtually all on a new Remington Selex G-R 4100, although other copiers have been found to be equivalent in performance so long as they have a setting for low intensity (paleness).

The platen glass was protected by a sheet(s) of A4 transparency. After selectively arranging the plant parts within an A4 format, on the folded newspaper, this clear sheet could be laid over them, giving a preview of what the photocopy would be like. The plant was then turned over onto the platen glass within this sandwich so as not to disturb the arrangement. The sheet also allowed the specimen to be slid about on the platen glass to permit ideal positioning.

An increased, larger than A4, approximately 5 mm thick pad of clean white tissue or paper towel was laid gently over the arrangement. This partly evened out the differences in thickness and excluded light from entering under the lid of the copier, and protected the lid from contact with the plant. The transparency and tissue paper were unchanged for over a hundred copies and had to be quite dirty before the photocopy was visibly affected. The edge of the transparency or the tape holding two transparencies together is visible as a line in some copies.

Style of layout is a personal one. Photocopy paper of A3 size would have been preferred over A4, as accommodating a far larger proportion of specimens without needing to cut them. However A4 was ruled to be the medium as being best for publication. It would be unfortunate if publication costs introduce any change in magnification from 1:1. Life size pictures lessen mental effort required when trying to find other character states to match specimens to the photocopy.

I tried to ensure no plant part extended to the edge of the A4 frame in any photocopy; perhaps wrongly, as several artists permit this to advantage (for example, T. Iwagu in Henty & Pritchard 1982, p. 144).

The machine operated well in general. Settings had to be much paler than for text, and could be made so by adjustment both on the console of the machine and on a higher level setting of 'exposure calibration' accessed by pressing the keys * 5 (as on a Canon machine). Alternatively, paleness was increased (it was discovered latterly) by setting the photo mode to on, and was used to remedy over-darkening when there was an abundance of toner in the machine. This mode appears not to otherwise improve resolution and seems to be at the cost of some sharpness. A dark copy has been accepted in preference for some specimens.

A blank longitudinal band somewhere down the page was corrected by removing the cleaning wire out of the machine and sliding the cleaner to park position. Bands of lighter and heavier intensity, not generally noticeable on copies of text, are seen on most copies of plants. They were partly eliminated if the keys * 6 were pressed to clean the rollers. Although they are aesthetically unpleasing, a combination of shades allows a single copy to display differences in detail of characters, such as venation and pubescence, which may be diagnostically important.

Two major difficulties were encountered with photocopying plant material. The parts cast a shadow of increasing darkness and extent according to their thickness and the darkness of the photocopier setting. Pressing helped flatten the specimens. Also excision or sectioning (which sometimes improved the botanical information of the photocopy) was sometimes required to reduce the thickness. It was still necessary to blank out the shadow on the majority of copies. This was done with typewriter fluid, which was often quite time consuming. Unfortunately, sometimes this created the possibility of introducing inaccuracies by the need for pencil or pen restoration of parts accidentally erased. Theoretically but not very practically it is possible to eliminate shadow by filling the spaces between thick parts with white tissue or, presumably, cotton wool, much as in a Reiker mount for insects.

Also, pale, particularly white or yellow, coloured parts failed to show, unless copied at a setting too dark to allow definition in the darker parts to be retained. Or they needed to be backed by a dark part such as a leaf, sometimes resulting in a rather artificial layout. An alternative was to make a very dark copy, using a setting which picks up at least the outline of the pale parts, and also make a second paler copy at a setting suitable for the dark parts. Then trace the pale parts onto the paler copy while it is laid over the darker copy, but with the risk of introducing error. Keeping pressed plants for some days until pale parts, such as white petals, had turned brown was another solution. The possibility of using a darkish paper as a background for pale specimens was inadequately explored.

A minimum of three photocopies were made of each plant. The worst one had written on it the specimen number, field notes, subsequent notes about the revisited plant or species, and determinations and literature references. The best was cleared of shadow and marks and awaits publication. The excess have on occasion provided replacements for lost and spoilt sheets and proved useful in providing a different view to serve identification. They will make two or three bound sets which will approach the prospectively published sets in botanical utility.

Technical inability resulted in there being a modal number of about five copies of each species; expending, very approximately: (769 + off-campus specimens + spoilt sheets) x ca. 5 = *ca.* 6 000 sheets.

A limitation throughout the work has been uncertainty about how the photocopies might be reproduced for publication. Certainly a photocopy of a photocopy loses too much of the detail seen on the specimen. Using typewriter fluid, or photocopying flowers and fruits on separate sheets, in the hope of later cutting and pasting them together, may turn out to be harmful to publication or redundant.

Presumably copies will go through the printer and gain a word- processed legend. Figure 1 is an example in not necessarily final form; with the specimen number, family, species, page (and figure) numbers in references, common name given by the latest reference, date and field data. The figure is from an untouched spare copy to give an indication of shadow and how the copy was rejected as too dark. Ironically this copy does not reveal the features (width of the bristle bases) needed to key this specimen to one of two species (Henty 1969, p. 46).

Identifications

I have perhaps tended to lump species, to produce conservative estimates of numbers of species. I did not attempt identifications of plants without reproductive parts and left several species unassigned to family rather than wrongly assign more than possibly one or two. The taxonomic arrangement as far as possible follows Höft (1992).

Categorization

Divisions between wild-versus-cultivated categories and life forms are based on an overall view of individuals of the species over the sites they occur in (ignoring seedlings). They are somewhat subjective, but probably not seriously so at the family level.

Obviously no plant would persist on campus unless it is being spared, at least temporarily, from the removal regime. In that sense all can be viewed as cultivated, or at least synanthropic. But cultivated plants is here intended for those which are present because humans restrain from removing or suppressing them, while they would others; or direct efforts at maintaining them, including importantly the removal or suppression of competing plants. This somewhat artificially places virtually all the trees in the cultivated category, but separating them would have been very subjective and arbitrary. The category includes indigenous, wild species brought, often by foresters, into the campus, or species indigenous to the campus which successfully benefit from similar protection. Wild vegetation is partly defined by it being subject rather to a regular removal regime, without selection among its plant species, and persistent growth in clearly untended, suitable sites. A further explanation of sources for identification and categorization will accompany the final analysis of the list.

RESULTS AND DISCUSSION

Photocopies

Photocopies can be nearly as good as herbarium specimens in showing grosser taxonomic features, even down to pubescence, and sometimes an improvement when they retain features seen only in the fresh specimen. Working with photocopies has been much more convenient in terms of weight and space (< 0.5 m^3 compared to >1.5 m^3 for herbarium sheets), convenience (I don't have to suffer air conditioning, nor exercise too much care with the annotated sheets or have to clean up bits of plant debris), and time (no duplicates have to be shuffled through and the copies can be kept to hand in home or office). Not unimportantly, they spare the herbarium specimens from wear due to repeated handling and do not have to be fumigated. They may sometimes permit the herbarium record to be dispensed with, resulting in significant savings in the time and expense sheets require in mounting and maintenance.

Numbers of species and families

Table 1 shows the 113 families which have been collected and the number of species within life forms within families within four higher taxa. Excluded from Table 1 (but included in the sum, last column) are seven wild shrubs (Caesalpiniaceae 2, Fabaceae 3, Malvaceae 1 and Solanaceae 1) and one wild tree (*Piper aduncum* L.). First exceptions to the view that all trees are cultivated would be some *Ficus* and other species allowed to persist, presumably for erosion control, alongside usually dry ditches, where several of the wild shrubs also grow. Noticeable is the differences in the length of the lists for the four higher taxa and the high number of grass species.

The 769 species so far recorded are enumerated in Table 2. The numbers of species of herbs is nearly twice that of trees; species of which number about three to four times as many as the number of each of the three remaining life forms, which have 173 species, fairly evenly divided among them.

This reflects the physiognomy of the campus and an administration policy of not permitting thick vegetation which can conceal wrongdoers. This is generally confined to hedges, thin borders along the sides of buildings and some peripheral areas such as the borrow pit.

Figure 1. An example of an untreated photocopy of a plant specimen (*Cenchrus* sp.), showing the types of shadow and marks that usually have to be removed and the type of legend which may be adopted.

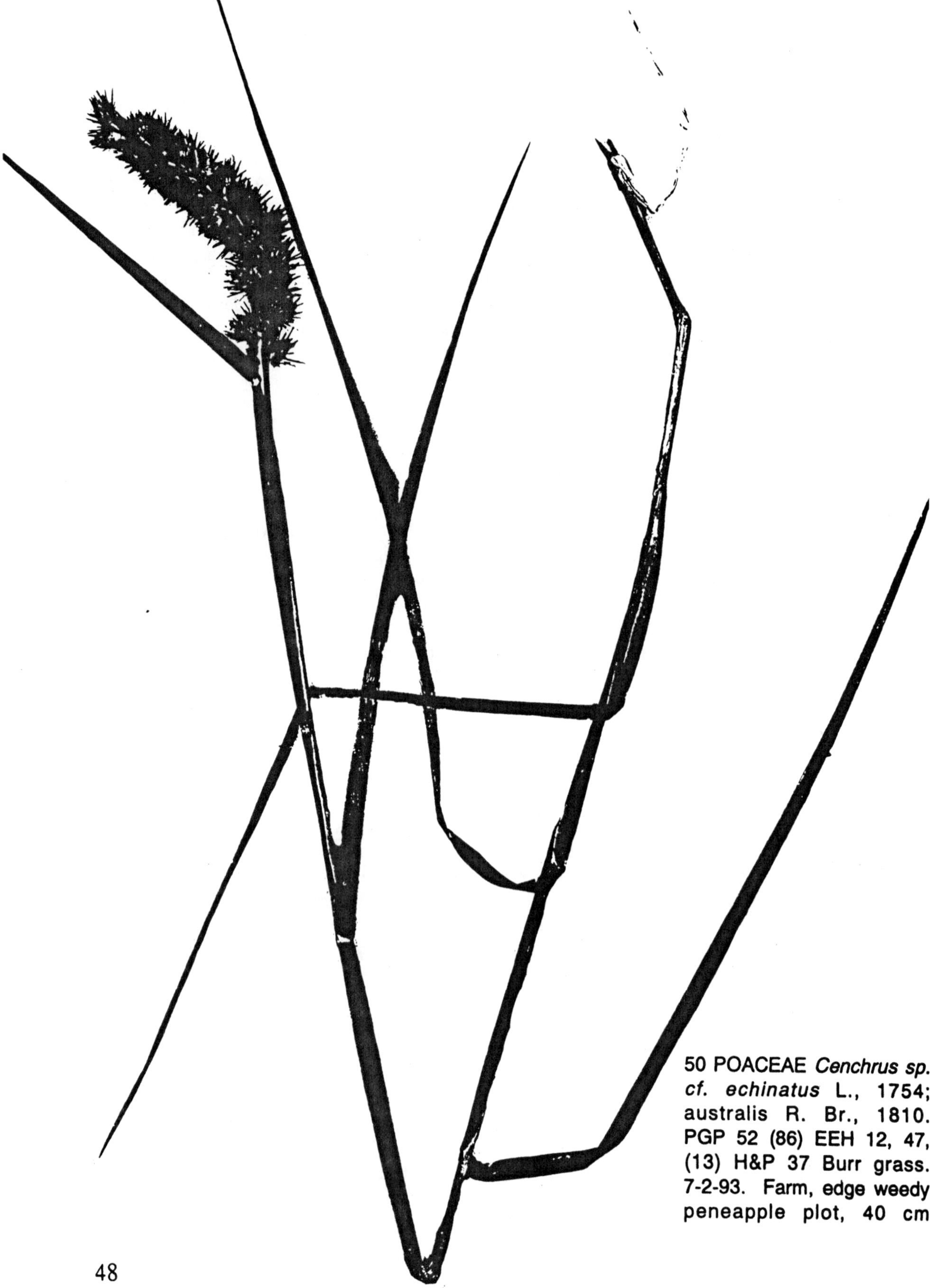

Mown grass with scattered trees, usually with their crowns not in contact, predominate on the common areas. Plenty of disturbed areas, the frequently burnt borrow pit and the farm provide habitat for ruderal species. Gardens are generally well tended but with opportunity for fast growing ruderals, usually have lawns, a lesser area of herbaceous beds and up to five to ten trees. A low hedge, typically of *Hibiscus rosa-sinensis* L., is often present but rarely complete enough to prevent access at several points. There are over twice as many cultivated (517) as wild (252) species (Table 2), although the difference is reduced when trees are ignored, leaving 309 and 251 species respectively, but partially restored if the 71 grass species (± 14 sedge species) are also excluded.

Although cultivated herb species do not greatly exceed wild ones, reflecting a rich ruderal flora, including a large proportion of Henty and Pritchard's (1982) weeds; the superiority in numbers of cultivated species is maintained within every life form except, noticeably, epiphytes. Epiphytes are not removed as a rule. The synusia is more predominant at Lae than at Calabar. There are more cultivated shrub species than cultivated epiphyte plus climber species (47), whereas there are nearly two times more wild epiphytic than wild climber plus shrub species (25).

There are over two times as many species of dicotyledons as monocotyledons, which are over four times as speciose as the pteridophytes. The eight gymnosperms are all cultivated trees with the exception of the one, so far distinguished, fertile cycad.

Distribution among higher taxa is rather different between the cultivated and wild categories. Monocotyledon species inferiority is far less in the wild flora, partly due to its numerous grass species, and because most of the trees are dicotyledonous. The wild flora also contains twice as many species of pteridophyte as the cultivated flora. Cultivated and wild dicotyledons exceed monocotolydons in numbers of species of climbers and shrubs but only one out of 17 cultivated epiphyte species are dicotyledons as opposed to eight out of 46 for wild epiphyte species. Only two of the 31 epiphytic pteridophyte species are recognized as cultivated.

The 113 families so far collected are enumerated in Table 2. Trends for families are broadly similar to those for species. Ten more contain herbaceous (64) than tree species, with the other three life forms together represented by species in 49 families.

Twice as many families contain cultivated compared to wild species. The cultivated species-containing families are about equal in herbs and trees. They have shrubs contained in more species than in climbers plus epiphytes (18). In contrast, in wild species-containing families epiphytic species numbers nearly equal climber plus shrub species numbers (12).

SUMMARY AND CONCLUSIONS

Photocopying plant specimens has a number of advantages over using herbarium sheets.

The UNITECH campus has a flora of predominantly cultivated species and many grasses, dominated by herbs and trees. Only in epiphytes are wild species more numerous. This reflects the highly anthropogenic nature of the habitat, the flora being taxonomically and ecologically very unlike the indigenous climax flora it has replaced. This is especially marked because not only are the cultivated species, particularly ornamentals and edible plants, almost all originally alien to Papua New Guinea, but so are most of the ruderal species. Reid (in prep.), in expanding on some ecological aspects, considers some of the reasons for this.

It is hoped comparable studies from other similar habitats will enable some generalizations about this type of environment, since it is a common one and likely to become more so, and readily available for pedagogic purposes and research.

ACKNOWLEDGEMENTS

I thank the numerous kind residents who have permitted observations and collections of specimens from their areas and the several people, unfortunately anonymous, who have climbed trees to collect for me.

I am very grateful to Mr. P. Kohun, Acting Head of Department of Agriculture, UNITECH, for assenting to the photocopier being used for plants and for the forbearance he and the other photocopier users in the Department have shown throughout this year.

Mr. T. Iwagu's excellent painstaking herbarium sheet preparation (from often rather poor material) and Mr. J. Simaga's identifications of trees and other extensive help, and the facilities of the Forestry Department, UNITECH, are greatly appreciated. I thank Dr. R. Höft for identification of ferns.

My appreciation goes to the staff of the Remington Company, Lae, who have been prompt, courteous and efficient on calls to maintain the photocopier.

REFERENCES

Gurnah, A. M. (1985). Effects of weed competition at different stages of growth on the yield of taro. *Field Crops Res.* **10**, 283-289.

Henty, E. E. (1969). *A Manual of the Grasses of New Guinea.* Department of Forests, Division of Botany, Lae, Bot. Bull. 1, 214 pp.

Henty, E.E. & Pritchard, G.H. (1982). *Weeds of New Guinea and Their Control.* Department of Forests, Division of Botany, Lae, Bot. Bull. 7, 186 pp.

Höft, R. (1992). *Plants of New Guinea and the Solomon Islands. Dictionary of the Genera and Families of Flowering Plants and Ferns.* Wau Ecology Institute Handbook No. 13, W.E.I., Wau, Papua New Guinea, ix + 169 pp.

Holm, L., Pancho, J. V., Herberger J. B. & Plucknett D. L. (1979). *A Geographical Atlas of World Weeds.* Wiley Interruptions, London, xii + 393 pp.

APPENDIX 1

Number of species of cultivated and wild plants within each of the life forms: herb (H), epiphyte (E), climber (C), shrub (S) or tree (T), within families within higher taxa; recorded from UNITECH campus, Lae, January to October 1993. See text for shrubs and trees categorized as wild.

Higher Taxon	Number of Species										
	Cultivated						Wild				All
	H	E	C	S	T	TOTAL	H	E	C	TOTAL	SUM
PTERIDOPHYTA											
Adiantaceae	2	0	0	0	0	2	1	0	0	1	3
Aspleniaceae	0	0	0	0	0	0	0	2	0	2	2
Cyatheaceae	0	0	0	0	1	1	0	0	0	0	1
Davalliaceae	3	0	0	0	0	3	3	3	0	6	9
Dennstaedtiaceae	1	0	0	0	0	1	0	0	0	0	1
Marattiaceae	0	0	0	1	0	1	0	0	0	0	1
Ophioglossaceae	0	0	0	0	0	0	0	1	0	1	1
Polypodiaceae	0	1	0	0	0	1	0	13	0	13	14
Schizaeaceae	2	0	0	0	0	2	0	1	0	1	3
Selaginellaceae	3	0	0	0	0	3	1	0	0	1	4
Thelypteridaceae	0	0	0	0	0	0	1	0	0	1	1
unidentified	3	1	0	0	0	4	1	9	0	10	14
GYMNOSPERMAE											
Araucariaceae	0	0	0	0	2	2	0	0	0	0	2
Cupressaceae	0	0	0	0	1	1	0	0	0	0	1
Cycadaceae	0	0	0	1	0	1	0	0	0	0	1
Gnetaceae	0	0	0	0	1	1	0	0	0	0	1
Pinaceae	0	0	0	0	2	2	0	0	0	0	2
Podocarpaceae	0	0	0	0	1	1	0	0	0	0	1
DICOTYLEDONAE											
Acanthaceae	9	0	2	3	9	23	5	0	0	5	28
Aizoaceae	0	0	0	0	0	0	1	0	0	1	1
Amaranthaceae	4	0	0	0	0	4	6	0	0	6	10
Anacardiaceae	0	0	0	0	5	5	0	0	0	0	5
Annonaceae	0	0	0	0	3	3	0	0	0	0	3
Apiaceae[1]	3	0	0	0	0	3	0	0	0	0	3
Apocynaceae	1	0	0	1	9	11	0	0	0	0	11
Araliaceae	0	0	0	2	1	3	0	0	0	0	3
Aristolochiaceae	0	0	2	0	0	2	1	0	0	1	3
Asclepiadaceae	1	0	0	1	0	2	0	5	0	5	7
Asteraceae	13	0	0	1	0	14	15	0	0	15	29
Balsaminaceae	1	0	0	0	0	1	0	0	0	0	1
Barringtoniaceae	0	0	0	0	6	6	0	0	0	0	6
Begoniaceae	6	0	0	0	0	6	0	0	0	0	6
Bignoniaceae	0	0	0	0	5	5	0	0	0	0	5
Bixaceae	0	0	0	0	3	3	0	0	0	0	3
Bombaceae	0	0	0	0	1	1	0	0	0	0	1
Brassicaceae[1]	3	0	0	0	0	3	0	0	0	0	3

Burseraceae	0	0	0	0	3	3	0	0	0	0	3
Boraginaceae	0	0	0	0	1	1	0	0	0	0	1
Cactaceae	3	0	0	0	0	3	0	0	0	0	3
Caesalpiniaceae	0	0	0	1	14	15	0	0	0	2	17
Capparaceae	1	0	0	0	0	1	2	0	0	2	3
Caricaceae	0	0	0	0	1	1	0	0	0	0	1
Casuarinaceae	0	0	0	0	2	2	0	0	0	0	2
Celastraceae	0	0	0	0	5	5	0	0	0	0	5
Combretaceae	0	0	0	0	9	9	0	0	0	0	9
Convolvulaceae[2]	0	0	5	0	0	5	0	0	6	6	11
Crassulaceae	3	0	0	0	0	3	0	0	0	0	3
Cucurbitaceae	0	0	4	0	0	4	0	0	2	2	6
Dilleniaceae	0	0	0	0	2	2	0	0	0	0	2
Elaeocarpaceae	0	0	0	0	1	1	0	0	0	0	1
Euphorbiaceae	3	0	0	6	11	20	7	0	0	7	27
Fabaceae	2	0	5	0	2	9	16	0	5	24	33
Flacourtiaceae	0	0	0	0	1	1	0	0	0	0	1
Gentianaceae[1]	1	0	0	0	0	1	0	0	0	0	1
Gesneriaceae	6	0	0	0	0	6	0	0	0	0	6
Goodeniaceae	0	0	0	0	1	1	0	0	0	0	1
Lamiaceae	7	0	0	1	0	8	7	0	0	7	15
Lauraceae	0	0	0	0	1	1	0	0	0	0	1
Leeaceae	0	0	0	0	1	1	0	0	0	0	1
Loranthaceae[2]	0	0	0	0	0	0	0	2	0	2	2
Lythraceae	0	0	0	1	1	2	0	0	0	0	2
Magnoliaceae	0	0	0	0	1	1	0	0	0	0	1
Malvaceae	3	0	0	6	2	11	4	0	0	5	16
Melastomataceae	2	0	0	0	0	2	0	0	0	0	2
Meliaceae	0	0	0	0	3	3	0	0	0	0	3
Mimosaceae	0	0	0	0	9	9	1	0	1	2	11
Araceae	0	0	0	0	16	1	6	0	0 0	0	16
Myrtaceae	0	0	0	0	7	7	0	0	0	0	7
Nepenthaceae	1	0	0	0	0	1	0	0	0	0	1
Nyctaginaceae	0	0	0	1	0	1	1	0	0	1	2
Nymphaeaceae[1]	1	0	0	0	0	1	0	0	0	0	1
Onagraceae	0	0	0	0	0	0	2	0	0	2	2
Oxalidaceae	0	0	0	0	1	1	1	0	0	1	2
Passifloraceae	0	0	1	0	0	1	0	0	1	1	2
Piperaceae	2	0	2	1	0	5	1	0	0	2	7
Polygalaceae	0	0	0	0	0	0	1	0	0	1	1
Polygonaceae	1	0	1	0	0	2	1	0	0	1	3
Portulacaceae	3	0	0	0	0	3	1	0	0	1	4
Proteaceae	0	0	0	0	1	1	0	0	0	0	1
Rosaceae	1	0	0	0	2	3	0	0	0	0	3
Rubiaceae[3]	0	1	0	11	3	15	4	1	0	5	20
Rutaceae	0	0	0	0	4	4	0	0	0	0	4
Sapindaceae	0	0	0	0	5	5	0	0	0	0	5
Scrophulariaceae	1	0	0	1	0	2	3	0	0	3	5
Simaroubaceae	0	0	0	0	1	1	0	0	0	0	1
Solanaceae	5	0	1	0	0	6	1	0	0	2	8
Sterculiaceae	0	0	0	0	5	5	1	0	0	1	6
Tiliaceae	1	0	0	0	1	2	1	0	0	1	3

Urticaceae	4	0	0	0	0	4	2	0	0	2	6
Verbenaceae	1	0	0	2	3	6	2	0	0	2	8
Vitaceae	0	0	1	0	0	1	0	0	1	1	2
unidentified	10	0	3	10	15	38	7	0	0	7	45
MONOCOTOLYDONAE											
Agavaceae	2	0	0	2	0	4	0	0	0	0	4
Aloeaceae	1	0	0	0	0	1	0	0	0	0	1
Amaryllidaceae	13	0	0	0	0	13	0	0	0	0	13
Araceae[1]	21	2	0	0	0	23	1	2	0	3	26
Arecaceae	0	0	0	0	18	18	0	0	1	1	19
Bromeliaceae	3	0	0	0	0	3	0	0	0	0	3
Cannaceae	2	0	0	0	0	2	0	0	0	0	2
Commelinaceae	4	0	0	0	0	4	3	0	0	3	7
Cyperaceae	1	0	0	0	0	1	13	0	0	13	14
Dioscoreaceae	0	0	2	0	0	2	0	0	1	1	3
Dracaenaceae	0	0	0	1	1	2	0	0	0	0	2
Iridaceae	2	0	0	0	0	2	0	0	0	0	2
Liliaceae	3	0	1	0	1	5	1	0	0	1	6
Marantaceae	4	0	0	0	0	4	1	0	0	1	5
Musaceae	6	0	0	0	0	6	0	0	0	0	6
Orchidaceae	2	12	0	0	0	14	0	7	0	7	21
Pandanaceae	0	0	0	0	2	2	0	0	0	0	2
Poaceae	10	0	0	1	1	12	59	0	0	59	71
Sansevieraceae	1	0	0	0	0	1	0	0	0	0	1
Zingiberaceae	14	0	0	0	0	14	1	0	0	1	15
unidentified	1	0	0	0	0	1	0	0	0	0	1

[1] Cultivated aquatics included: *Pistia stratiotes* L. in Araceae, one marketed edible highland species introduced in Apiaceae, one *Rorippa* in Brassicaceae, *Nymphoides indica* (L.) Thw. in Gentianaceae, *Nymphaea* in Nymphaceae, one elodeaoid in unidentified Dicotyledonae.

[2] Wild parasites included: *Cuscuta australis* R. Br., in Convolvulaceae, and both Loranthaceae.

[3] Ant plants included: one wild, one cultivated.

APPENDIX 2

Number of species and familes of cultivated and wild plants within each of the life forms: herb (H), epiphyte (E), climber (C), shrub (S) or tree (T); and, for species, within higher taxa: recorded from UNITECH campus, Lae, January to October 1993.

HIGHER TAXON	H	E	C	S	T	TOTAL
			NUMBER OF SPECIES			
			CULTIVATED			
PTERIDOPHYTA	14	2	0	1	1	18
GYMNOSPERMAE	0	0	0	1	7	8
DICOTYLEDONAE	103	1	27	49	177	357
MONOCOTOLYDONAE	90	14	3	4	23	134
ALL TAXA	207	17	30	55	208	517
			WILD			
PTERIDOPHYTA	7	29	0	0	0	36
GYMNOSPERMAE	0	0	0	0	0	0
DICOTYLEDONAE	94	8	16	7	1	126
MONOCOTOLYDONAE	79	9	2	0	0	90
ALL TAXA	180	46	18	7	1	252
			CULTIVATED AND WILD			
PTERIDOPHYTA	21	31	0	1	1	54
GYMNOSPERMAE	0	0	0	1	7	8
DICOTYLEDONAE	197	9	43	56	178	483
MONOCOTOLYDONAE	169	23	5	4	23	224
ALL TAXA	387	63	48	62	209	769
NUMBER OF FAMILIES	H	E	C	S	T	TOTAL
			CULTIVATED			
ALL TAXA	54	5	13	21	53	106
			WILD			
ALL TAXA	38	11	8	4	1	52
			CULTIVATED AND WILD			
ALL TAXA	64	11	15	23	54	113

SESSION II: VARIATION IN FORESTRY AND AGRUCULTURE

Proceedings of the Biological Society of New Guinea
R. Höft (ed.)

AN EXPLORATION OF CULTIVATED BANANA CULTIVARS IN YOOM-NUNI VILLAGE OF MANOKWARI DISTRICT, IRIAN JAYA

Maria Justina Sadsoetoeboen & A. Kilmaskossu

Faperta UNCEN Manokwari, P.O. Box 23, Manokwari 98314, Indonesia

The objective of the research was to study the number and variation of morphological characteristics of cultivated banana in Yoom-Nuni village in the Manokwari district of Irian Jaya. The descriptive methods used in collecting the data were interviews and direct observation of the gardens.

The results of the study showed that there were 14 cultivars of cultivated bananas that could be clearly distinguished based on the variation of their fruits. The cultivars, however, could not be clearly distinguished based on the variation of three other characters which were pseudo-stem, petiole margin and bracts.

Based on the way the local people utilize the banana as food, the 14 cultivars of cultivated banana could be divided into two groups. The first group included four cultivars that could be eaten directly, and the second group included 10 cultivars that could be eaten after cooking.

Proceedings of the Biological Society
R. Höft (ed.)

VARIATION WITHIN *POMETIA* (SAPINDACEAE) SPECIES IN PAPUA NEW GUINEA

Kipiro Damas

Botany Branch, Papua New Guinea Forest Research Institute, P.O. Box 314, Lae, Papua New Guinea

INTRODUCTION

Taxonomic History of Pometia

The classification of the genus *Pometia* into highly distinguished nomenclature (species) has been a problem for quite some time because of its variability. Two comprehensive studies of the genus have been made. The first by Blume (1847), where the genus was named *Iriana*, was an admirable piece of early descriptive work. The second study was published by Radlkofer (1933). Here the sub-divisions of the genus were made but all the groups maintained in the rank of species. In 1855, the first Botanist to recognize a singular nervation in *Pometia* was Thwaites in his excellent description of *Ecremanthus* (= *Pometia*), a fact being mentioned by Hiern (1875).

The character disappeared completely from phytography until Endert (1928) mentioned it again. It gives an impression that vegetative characters for taxonomic distinction have been neglected in favour of reproductive characters. This may well be due to the influence of the Linnaeus Sexual System.

Jacobs (1962), studying the variability in the genus, mentioned that even placing the genus into varieties would be too high a rank for them; therefore they have been given the rank of forma.

Background of the study

In the past the genus *Pometia* in the family Sapindaceae has caused much confusion among botanists and others. The genus is known only by two species, *Pometia pinnata* and *Pometia ridleyi*. *P. ridleyi* is distributed throughout Malaysia except Singapore. *P. pinnata* is known by 8 forms and is found in Ceylon, the Andamans and throughout Malesia to Samoa (compare Figure 1). It is also found in a few scattered areas in North Siam, South Yunan, Indo-China and Formosa.

The species *P. pinnata*, having several different varieties, was subdivided into formae (forms). In Papua New Guinea the species is known by 4 out of the 8 recorded forms. These are f. *pinnata*, f. *glabra*, f. *repanda* and f. *tomentosa*.

Some of these forms are known to be of commercial importance to the region in terms of timber production. Apart from this, the fruits of some forms are eaten (see also Tanjung, this volume) depending on the thickness of the aril (flesh). At present, two of the forms of *P. pinnata* are among the major timber export species in Papua New Guinea. Both these forms are classified under a single name, *Pometia pinnata* even though they have strikingly differing characters.

The loggers in most parts of the country have always argued that they should be separated on species level and one should be known as *P. pinnata* and the other as *P. tomentosa*. It is for this reason that this project was undertaken. It is particularly aimed at seeing if there is any real differences between the two forms *P. pinnata* f. *pinnata* and *P. pinnata* f. *tomentosa*.

Figure 1. The larger plant-geographical regions

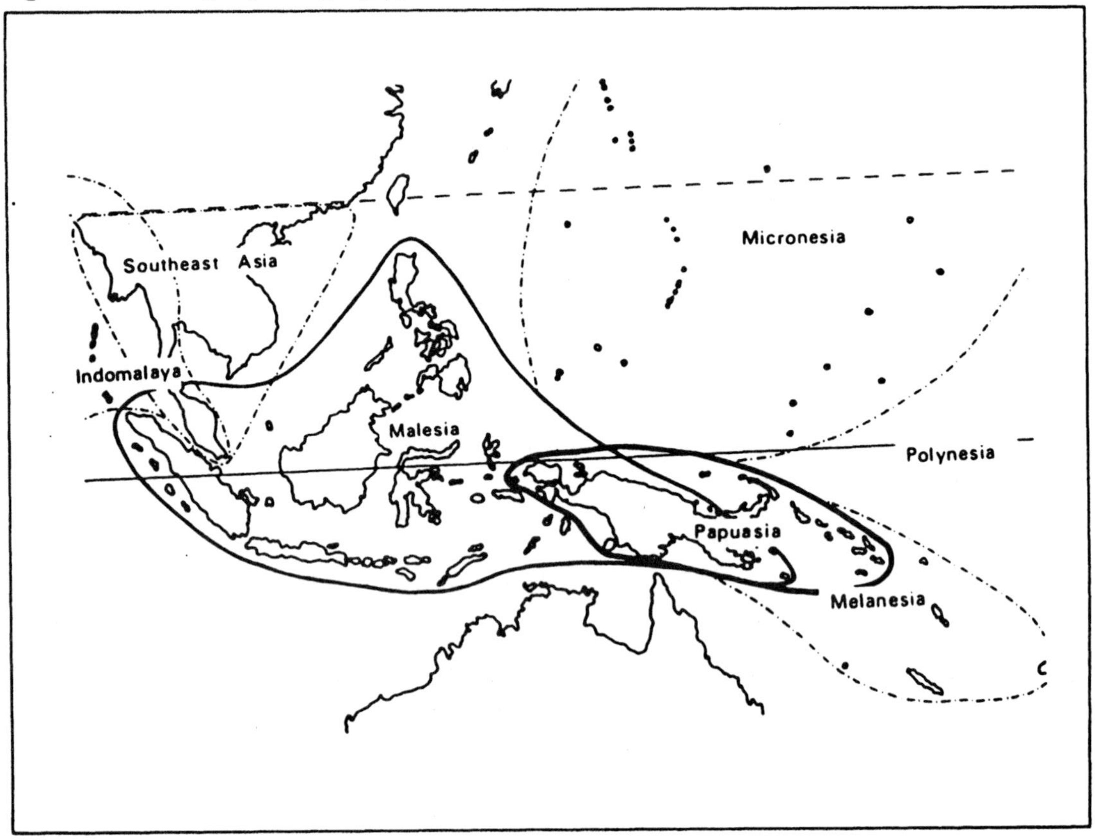

METHODS OF DESCRIPTION

Three different methods were used in searching for variations within the forms. That is:

- taxonomically, by analysing herbarium specimens and material collected by the author;
- anatomically, by analysing the wood structure of collections from various localities, and
- mophologically, by growing seeds of different forms and studying the seedlings, and by field observations of some saplings.

Detailed descriptions of methods will be given under each study.

Two forms (*P .pinnata* f. *pinnata* and *P. pinnata* f. *tomentosa*, being termed *P. pinnata* and *P. tomentosa*, respectively in this paper) were selected for the study being of particular interest to loggers and foresters.

One group of samples under *P. tomentosa* was based on specimens gathered from Lae, Madang and the New Guinea Islands and the other under *P. pinnata* was widely distributed. Taxonomically and morphologically, the collections were grouped according to leaf characters. An analysis of the floral parts was considered to be too time-consuming and not much floral variation was observed.

Taxonomic Observations

Specimens from the National Herbarium, Lae and my own collections from various places were analysed. Specimens were placed together according to their appearances using leaf characters, *i.e.* shape, size, colour, texture etc. Basically, the procedure of taxonomic revisions was followed, including measurement of leaves, inflorescence etc.

The following observations were made from the specimens:

- Specimen number, for herbarium specimens
- Locality including date of collection and altitude

- Leaf characters including numbers of leaflets per rachis leaf lenght, leaflet dimensions shape, texture (pubescent or glabrous) and lustre.
- Floral characters including spike lenght, number of branches, lustre (Note: flowers were not looked at in detail because generally most of them were alike)
- Fruit characters (if fruits were present) including shape and dimensions.

DESCRIPTION OF THE GENUS

Pometia J.R. & G. Forst.

Pometia includes trees ataining big sizes often with large buttresses up to 1.5 m or more. Often the juvenile shoots or innovations have a bright reddish or purplish tinge, (hirsute) with distinct rather stiff or bristly hairs. Leaves compound, paripinnate up to 1 m long with 4 to 17 or more pairs of leaflets. Leaflets coriaceous to herbaceous, often increased in size, on average 12 to 30 x 4 to 10 cm, sometimes larger; venation distinct with midribs and veins more prominent beneath; blade assymetrical, the acroscopical half being wider and more extended at the base. Inflorescence terminal or rarely axillary. Flowers polygamous. Male flowers about 4 mm across; calyx cup-shaped with 5-8 lobes, with erect teeth; petals 5, rarely as long as the calyx lobes; disc thick, yellow, lacking ovary; stamens 5-8, early stage is sessile, later long exerted, about 5-6 mm long. Female flowers about 6-7 mm; calyx lobes 5, petals 5, roundish, slightly exceeding the calyx lobes and wider than these, white; disc thick, yellow; ovary on disc 2-3 lobed surrounded by 5-8 staminodes; style 1 in the centre of the ovary lobes, 5 mm long; stigma 2. Fruit 2-lobed, only one develops (rarely 3) to egg shape, up to 5.0 x 3.0 cm, some are edible depending on thickness of the flesh (aril) covering the seed.

Distribution: *Pometia* is a pantropic genus. Its distribution extends from Ceylon, the Andamans throughout Malesia to Samoa. Few scattered in N. Siam, S. Yunan, Indo-China, and Formosa.

Ecology: Typical rain forest genus of low altitude, generally occurring below 500 m, rarely to 1,000 m. (highest record: 1,700 m in Atej) on limestone or loamy soils. Not so dominant in the forest in West Malaysia; in Malaya mainly riverine; in Borneo and Sumatra occasionally in fresh water swamp forests; in New Guinea not seldom dominant in forests partly under human influence.

Notes : The genus is recognized by two species, *i.e. P. pinnata* and *P. ridleyi*.

Pometia is well-known for its witches-brooms which occur nearly in all the taxa (species and forms) that can be recognized from long distances. Witches-brooms originate from a leaflet (Bos 1957). The malformation is caused by viruses. This characteristic also appears on inflorescences.

Field Key to the Forms

1. Tree, leaves compound paripinnate up to 1 m or longer; leaflets coriaceous to herbaceous.
   a. Leaflets assymetrical, sometimes ovate to lanceolate; base obliquely cordate, adaxial section overlapping the rachis.......... *P. pinnata* f. *tomentosa*
   b. Leaflets symetrical, oblong to elliptic rarely ovate; base shallowly cordate, rarely or not overlapping the rachis............ *P. pinnata*

Figure 2. Distribution of Pometia in Papua New Guinea

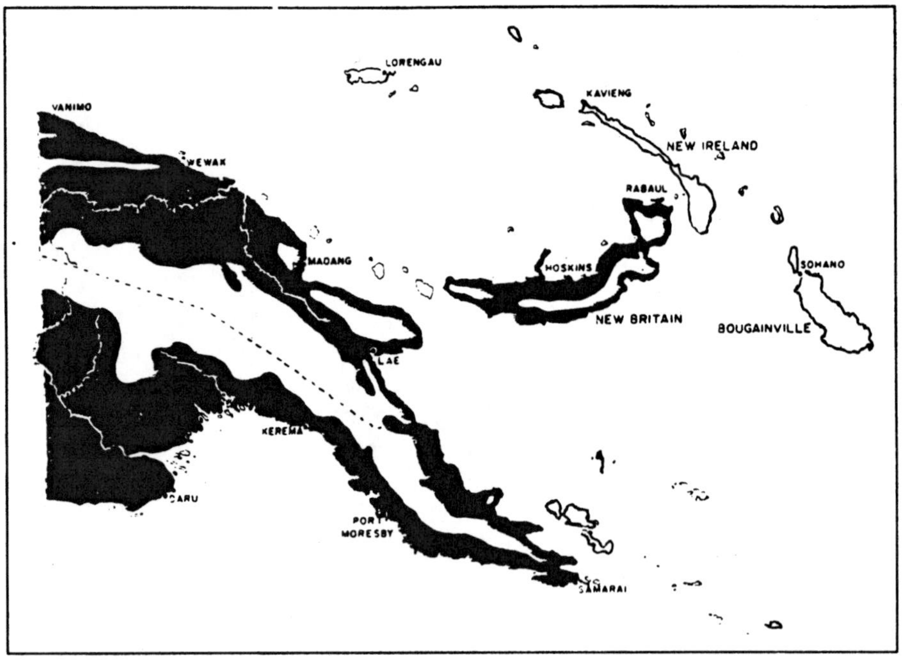

Pometia pinnata forma *tomentosa* (Bl.) Jacobs

Tree, 30-40 m tall; twigs terete, rusty-brown hairy; leaves 20-30 cm long; leaflets 5-7 pairs per rachis; blade falcate, assymetrical, sometimes ovate to lanceolate, 15-25 x 6-9 cm; mature leaflets glabrous, juvenile leaflets pubescent; margin toothed but sometimes faintly; base obliquely cordate, adaxial section overlapping the rachis; apex acute; inflorescence densely pubescent, 20-40 cm long; fruit 3-3.5 x 2-2.5 cm, sub-globose or ellipsoidal, red or brown when ripe; seeds oblong; testa reddish.

Field characters: Buttresses bluntly flanged, rarely up to 2 m or more; usually straight, rounded bole up to 30 m before the first limb; outer bark light brown, rough surface; branching less dense but usually heavily foliaged, foliage lighter or somewhat yellowish green.

Distribution (Based on the material studied): Morobe , Madang, East and West New Britain, New Ireland Provinces and Bougainville.

Ecology: Primary lowland rain forest on alluvial flats or foothills and on hilly slopes up to 200 m.

Specimens cited: Kerema. Schodde R. 2470; Brass 27573, 29209, Schodde & Craven 4081; NGF, 3343, 3273, 11182, 216, 32253, 10658, 6565, 154559, 16413, 31639, 1695, 6444, 36085, 65651, 21900.

Pometia pinnata Forst.

Tree, 20-30 m tall; twigs terete, hairy, when older completely glabrous or almost so; leaves 30-40 cm long (up to 100 cm); leaflets coriaceous, 10-11 pairs, fairly symetrical, oblong to elliptic, rarely ovate, 20-25 x 7-8 cm, first pair usually sub-orbicular to elliptic, clasping the rachis-like stipules; base oblique, shallowly cordate; tip acute, sometimes acuminate; inflorescence terminal in glabrous or finely pubescent panicles, 30-50 cm long; fruit lobes 5-6 cm across, flattened to globose, usually only one lobe is fully developed, 1.5-3.5 x 1-3 cm; exocarp leathery, inner fibrous-meaty, white or reddish; aril fleshy, white, pleasant tasting; seed irregular red-brown with a large hilum.

Field characters: Heavily buttressed, thin or sharply flanged, up to 3 m or more; bole often not so straight, 50-60 cm diameter, rarely rounded reaching up to 20 m before the first limb; bark more or less smooth, dark brown; branching low with heavy dark green foliage.

Distribution: Same as that of the genus. (In Papua New Guinea occurring in almost all coastal regions.)

Ecology: Predominent in primary forest on all kinds of soil, river banks or well drained limestone etc.

Specimens cited: Hartley 10074A; Pullen 7556; Artis Vinas 33; Saunders J. S. 559, 5048, NGF 7444, 1407, 1695, 14637, 1549, 46556, 46590, 15461, 42133.

MORPHOLOGICAL DESCRIPTION

Methods

Seeds of different forms were collected and germinated. Description of seedlings was made in the first stage after germination and in the second stage after cotyledons had fallen off:

The following features were recorded:

- rooting behaviour;
- hypogenic or epigenic development of seedling;
- thickness of collet;
- hypocotyle: lenght and lustre;
- cotyledon: size, shape, lustre, colour;
- epicotyle: lenght, shape, lustre;
- first leaf: colour, lustre, leaflet size, shape;
- second leaf: colour, lustre, leaflet size, shape;
- general apperance of the plant.

Field Key to the Seedlings

1. First two leaves compound, imparipinnate opposite or sub-opposite.
   a. Leaflets slightly glossy light green above, paler beneath, hairy like stems especially on the nerves .... *P. pinnata* f. *tomentosa*
   b. Leaflets glossy red-brownish green above, paler beneath glabrous .... *P. pinnata*

Seedling description of *Pometia pinnata* forma *tomentosa* (Bl.) Jacobs.

Roots brown, turning reddish-yellowish; tips white; primary root thicker than the lateral ones, moderately flexuous, tapering downwards; collet slightly thickened; lateral roots very dense at the base, reducing downwards, very flexuous, much branched; hypocotyle epigeous, errect, straight, 8-9 cm long, 2-3 mm thick, terete, glabrous, with fine red streaks. (turning brown); cotyledons epigeous, 2, opposite, unequeal, sessile, usually largest one curved upwards and smaller downwards or straight; blade obovate, long with index 3-4 x 1-1.5 cm, 7-8 mm thick, thick and fleshy, flat or concave above, sort of globose or concave beneath, having acute top and an amplexicaul (stem-clasping) base, indistinctly nerved, brownish green above, brown beneath somtimes red-blotched, glabrous. Epicotyle erect, obtuse-angular beneath the leaves, finely hairy. Phyllotaxy spiral but first two leaves opposite or sub-opposite. Leaves exstipulate, paripinnate, but first leaf imparipinnate having 2 pairs of lateral leaflets and a terminal; higher leaves with more pairs; rachis thickened at the base, green, hairy like the epicotyle; leaflets opposite or sub-opposite, petiolulate, first leaf sessile except terminal leaflet; petiolule 3-4 mm long, rounded; green, hairy, blades ovate or ovate-oblong or ovate-lanceolate; lateral leaflets often or less falcate, thin, having obtuse or acuminate tip, pinninerved; slightly prominently nerved on both surfaces; margin dentate, slightly glossy, light green above, dull green beneath, on both surface, especially on the nerves, hairy like the stem; first leaf rachis short, 2.5-3.5 cm; blades in terminal leaflets usually largest; succeeding leaves paripinnate, mucronate; top of the plant light yellowish green, densely hairy.

(Note: Seeds from two mother plants were germinated and described)

Seedling description of *Pometia pinnata* Forst.

Roots brown; primary root flexuous, much thicker that lateral ones; collet not thickened; hypocotyle epigeous, erect, straight 10 cm long, red-pink in early germination stage, very finely hairy; cotyledons epigeous, 2, fleshy, opposite brown-reddish, flattened or slightly concave above, 3-3.5 cm long, 1-3 cm at the widest. Epicotyle erect, straight, 18 cm long, smaller than the hypocotyle, 8 cm long, hairy like hypocotyle. Phyllotaxy spiral but first 2 leaves imparipinnate, 2 pairs of lateral leaflets and a terminal one; terminal leaflets largest, lanceolate, gradually changing to elliptic or obovate; rachis 2.5-3.0 cm long, 1.5-2.0 mm thick, thickened at the base, slightly flattened above, convex beneath, finely hairy; leaflets opposite or subopposite, shortly petioluloate, lowest pair smaller; blades elliptic or obovate, having acute to accuminate tip, pinninerved; midrib prominent on both surfaces but lateral nerves prominent beneath, hairy, like rachis; margin dentate, glossy red-brownish green above, paler beneath, later turning darker green above, paler beneath. Succeeding leaves paripinnate, mucronate. Top of the plant glabrous green.

(Seeds from two mother plants)

Anatomical Description

Pometia pinnata f. *tomentosa* (Bl.) Jacobs

Wood (heart wood) straw to light brown; pores intermediate, in short radial multiples; parenchyma surrounding the pores and in irregular bands. (Appendix 5)

Pometia pinnata Forst.

Wood (heart wood) dark reddish brown; pores intermediate, in short radial multiples; rays narrow; parenchyma surrounding the pores. (Appendix 6)

DISCUSSION

The groupings obtained give the impression that the form *tomentosa* should be a species of its own. The hierachy seems to be: Species - sub-species, variety, sub-variety, forma and clone. So even if it is called a variety it would be a promotion. Certainly this won't be the case as yet because of several complications I have encountered. These are yet to be solved.

Taxonomical Groupings

While taxonomically grouping specimens into their respective resemblance groups, it was seen that some of the specimens possessed completely different appearances. There seemed to be a trend following the geographical localities. The trend which is to be proven is that species known by a certain name in one locality is different from the same species known in a different locality.

A good example noted was of the specimen collected from Bulolo which is known as *P. tomentosa*. It has slightly different leaflet characters from that of the same species known from arround Lae, Madang and New Britain areas (Figures 3 and 4). It was also noted that this specimen from Bulolo alone seems to match the early descriptions of *P. tomentosa*.

Figure 3. Illustration of *Pometia pinnata* forma *tomentosa* (A leaf arrangement, B flower, C stamen, D petal, E style)

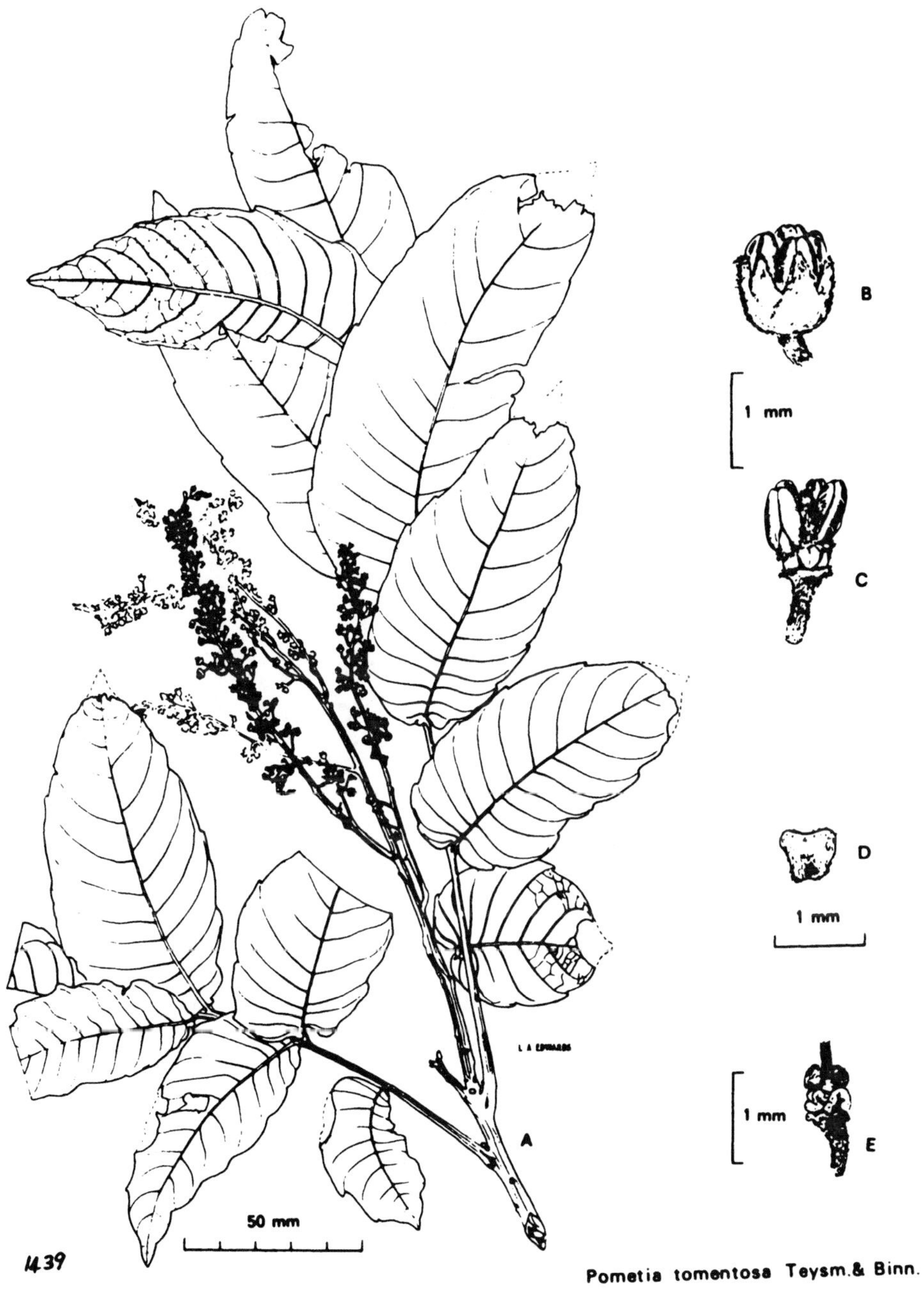

Figure 4. Illustration of *Pometia pinnata* forma tomentosa. Streiman's description based on this specimen from Bulolo

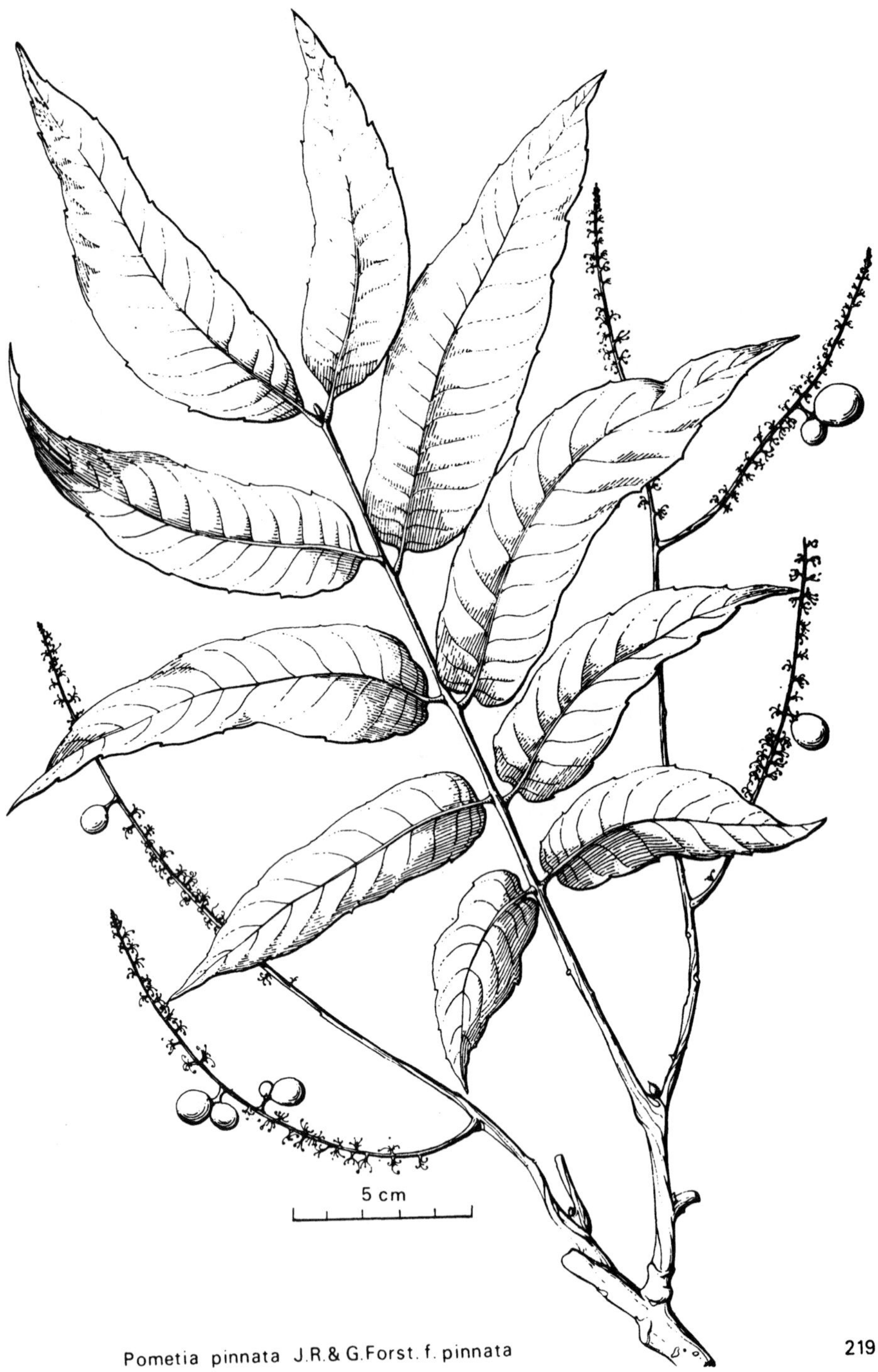

Figure 5. Illustration of *Pometia pinnata* (A leaf, B style, C stamen)

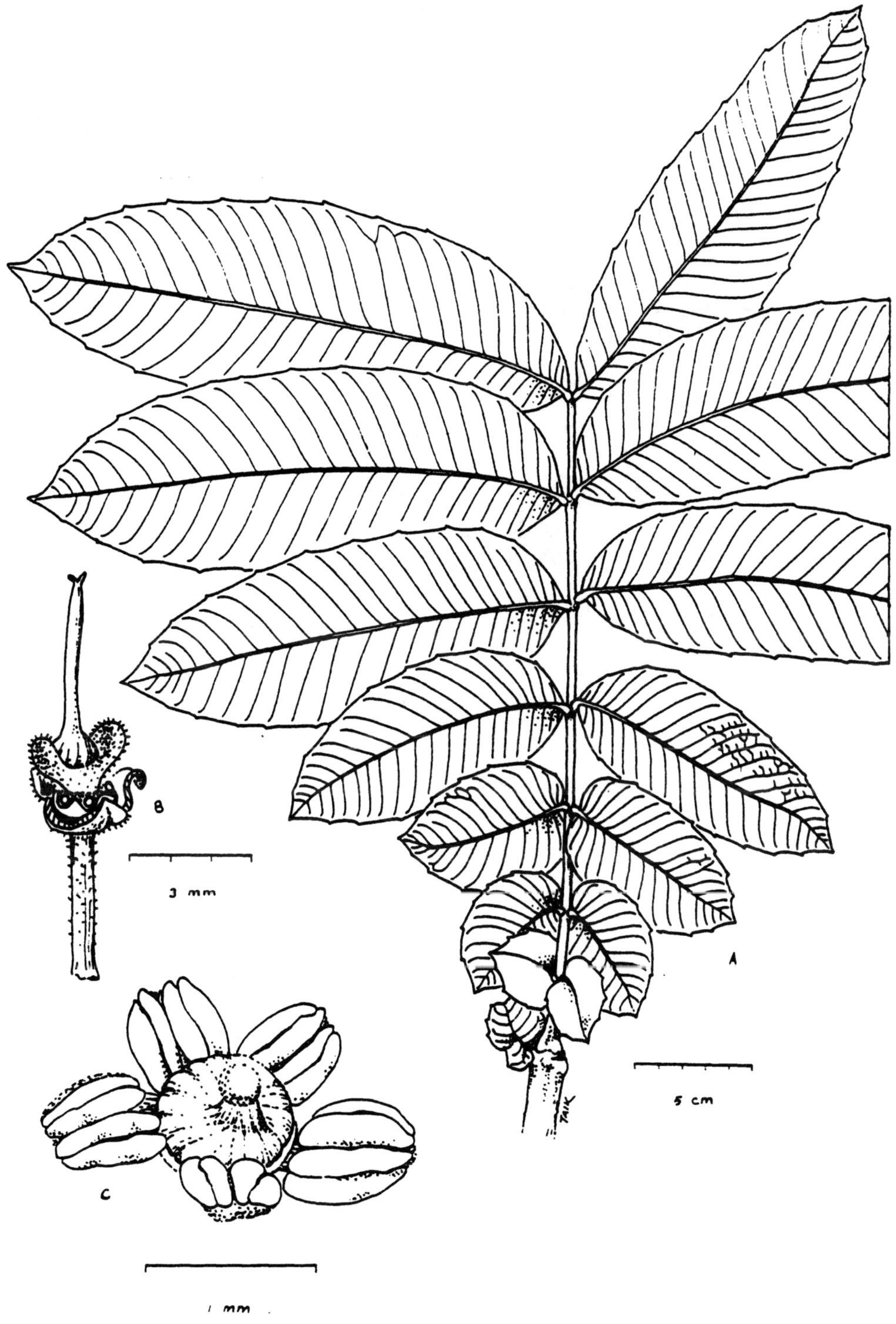

Figure 6. Illustration of seedling of *Pometia pinnata* forma *tomentosa*

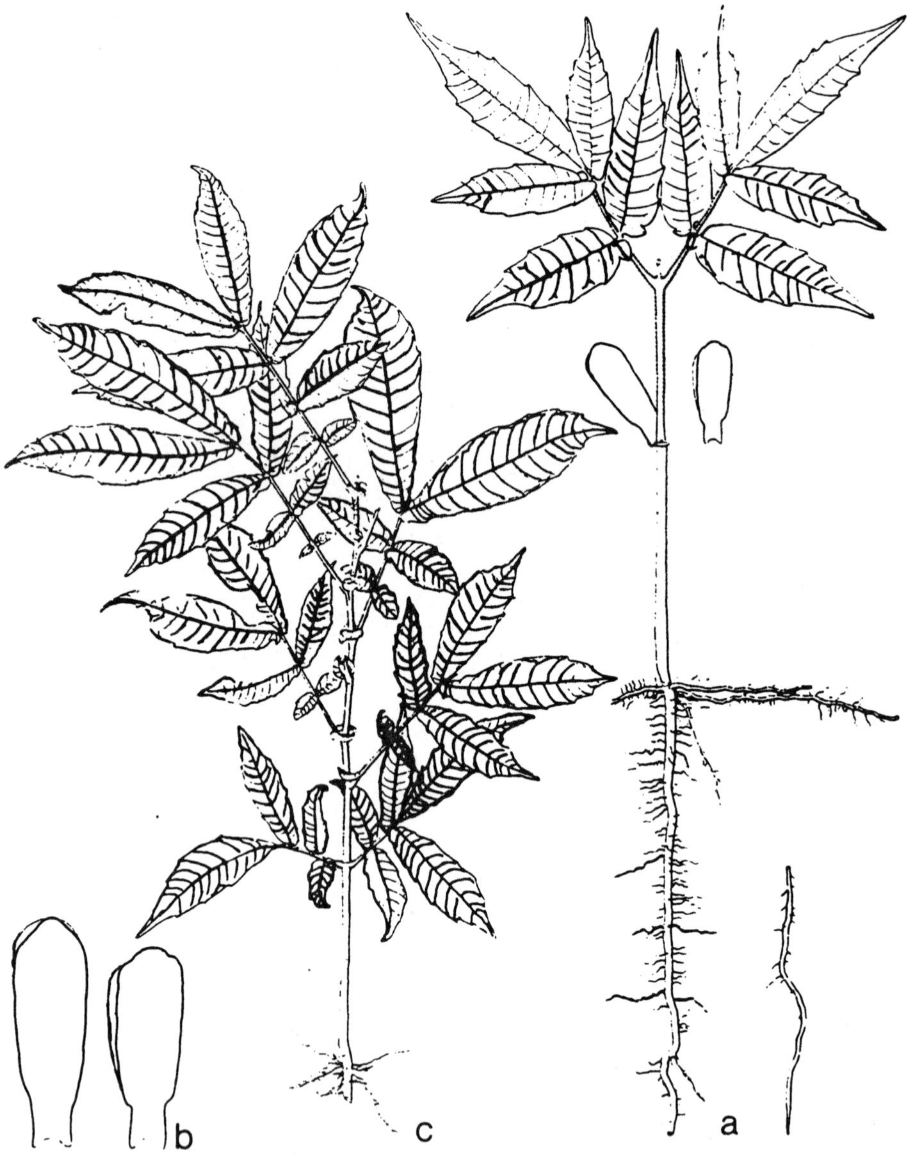

Figure 7. Illustration of seedling of Pometia pinnata with the first pair of leaves developed

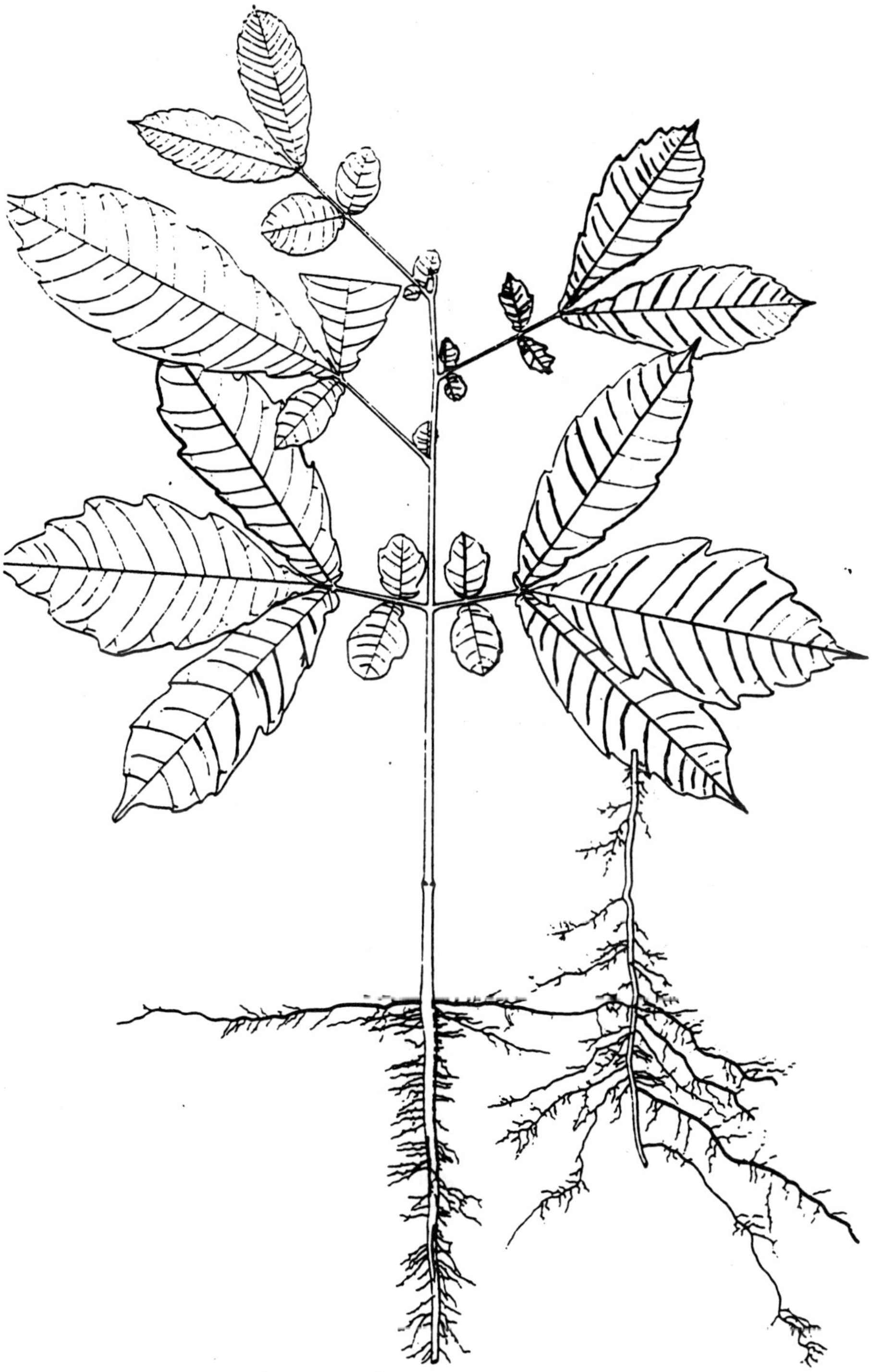

Forms Encountered

I have also encountered a couple of forms which did not seem to have any resemblance to *P. pinnata* and *P. tomentosa*. These specimens observed were from two different localities. One from New Ireland known as "wild taun" which did not seem to have been mentioned in the literature searched. It had very narrow, elliptical, and glabrous mature leaflets. Only the juvenile leaflets matched that of *P. tomentosa*. The other form from Northern Province (Kokoda trail) is known as *P. coriaceae*. It also had a rather peculiar appearance, having smaller leaflet sizes, with a distinctly toothed margin. Its base was tapering or narrowing into the petiolule. Specimens such as these were put to prevent complications.

So basically the description I have made here for *P. tomentosa* were based on specimens gathered from several different localities.

Morphology (Seedlings and Saplings)

Seedlings observed of both forms showed that earlier in their germination stage, both appeared to be similar in appearance, gradually changing as the seedlings grew older, *i.e.* after the fourth to fifth leaves had appeared. *P. pinnata* had glabrous and glossy leaflets while *P. tomentosa* still maintained its hairyness on twigs and leaflets.

The colour difference was evident with *P. tomentosa* being lighter green while *P. pinnata* had a much darker apperance. Also the size and shape of the leaflets varied in that *P. tomentosa* has much narrower leaflets than *P. pinnata* (Figure 3).

Field Observations

Other features in the field including buttresses, bark surfaces, colour and bole length, and shape have also been noted to vary. *P. pinnata* seems to have high and sharply flanged buttresses, the bark flaking off in short thin flakes leaving the surface smoother and is much darker in colour. It also has a shorter bole and an irregularly rounded shape.

P. tomentosa on the other hand is less buttressed or less in height and bluntly flanged and its external bark surface is somewhat finer (Fig 4 B.), peeling off in short thin flakes leaving distinct scars or more or less shallow craters. The colour is usually light to reddish brown. The bole is usually straight, longer and more or less nicely rounded, not geometrical but reasonable (Fig 4 A.).

Additional features in mature trees noted included branching habit and stuntedness of the species which also contribute to the variations. Here, *P. pinnata* seemed to have low branching and was almost often seen to be stunted in growth in the sense that it sometimes had twisted boles coppicing much lower down. *P. tomentosa* seemed to be free of such irregularities.

Wood Anatomy

From the point of view of wood anatomy, not enough observations have been made, but significant differences were seen in the samples observed. The cross sections of both forms were observed and the visible features seen were the parenchyma arrangements.

P. pinnata had vasicentric parenchyma and *P. pinnata* forma *tomentosa* had both vasicentric and irregular bands of parenchyma (Fig 5).

CONCLUSIONS

It seems that each of the methods used in an attempt to figure out variations within the forms have indicated that certainly there were some characters that confirm that the forms could not be known as one species or rather grouped under one name. Taxonomically, the variations were seen in the mature forms of the species, morphologically the variations in the juvenile stages of the species were seen and anatomically the internal wood structures were seen. All the illustrations shown do indicate obvious variations. *Pometia pinnata* can be distinguished from *Pometia pinnata* forma *tomentosa* at the late seedling stage, as saplings and when trees are mature.

However, a problem which needs to be clarified is the geographical variability. I have raised this point in my discussion because I have noted that a specimen known as *Pometia tomentosa* in Bulolo area is quite different from that same species known in Lae, Madang and the New Guinea Islands. Such cases may occur for different localities.

It is also noted that most of the literature sought does not specify the forms their descriptions relate to. An example in the description by Havel (1975) obviously relates to *Pometia pinnata* forma *tomentosa* and Streimann (1972) described the form that occurs in Bulolo and Gulf. These are probably higher altitude forms (according to me). The descriptions do not match the same form (f. *tomentosa*) known in other areas. However, in a footnote Streiman points out clearly that there is another form of *Pometia pinnata* which does not have the same features, rather an undesirable form.

Jacobs (1962) who did extensive studies on the variation within the genus *Pometia* pointed out with the limited material he had, that the groups would not deserve a separation on species level.

Observing such cases here, I feel quite dissatisfied considering my objectives. Especially knowing that this particular species in this particular region is very important economically. Having a reasonable number of characters to differentiate them, and seeing them grouped under one name is a bit unjustifiable.

I therefore recommend here that any future study should be based on Jacobs work and that the descriptions to be made must specify which particular form they relate to. In addition to this, we should know this species by its specific name locally. Anatomically , more work is required on strenght, durability etc. for each of the species or forms.

ACKNOWLEDGEMENTS

In a different form, this paper was presented as my major paper for my degree at Unitech in 1992, in which I acknowledged many who assisted me.
For this part, I would like to mention a few FRI Staff for their support in putting together this paper. Firstly to my Branch Heads (Botany) and the rest of the Botany Staff for their most required encouragements. In Forest Products Branch, special thanks to Anna Aglua for spending her time in the preparation of slides for my anatomical descriptions.

REFERENCES

Blume, C.L. (1847). Rumphia (3): 113-117.

Bos L. (1957). Meded. Landbouwog. Wangen. 57 (1): 1-79.

Burger, D. H. (1972). Seedlings of some tropical trees and shrubs mainly of South East Asia.

Corner, J. H. (1952). Wayside trees of Malaya Vol. 1.

Eddowes, P.J. (1977). Commercial Timbers of Papua New Guinea.

Endert (1928). Geslachtstabellen.

Havel, J.J. (1975). Training Manual For The Forestry College, Vol. 3 part 2.

Hiern, W.P. (1875). Flora of British India.

Jacobs, M. (1962). *Pometia* (Sapindaceae). A Study in Variability.

Peekel, P.G. (1984). Flora of Bismarck Archipelago. Reprint 1984

Radlkofer, L. (1933). Pflanzenreich, Heft 98: 924-936.

Streimann, H. (1972). *Pometia pinnata* forma *tomentosa* (*P. tomentosa*) Sapindaceae. Timber Species Leaflet No.7.

Thwaites, G.H.K (1855). Hook., J. Bot. and Kew Gard. Misc. 7: 272, t. 9.

Wyatt-Smith (1954). Manual of Malayan Timber Species no. 6.

APPENDICES

Appendix 1. Sapling of *Pometia pinnata* forma *tomentosa*

Appendix 2. Sapling of *Pometia pinnata*

Appendix 3. Mature tree of *Pometia pinnata* forma *tomentosa* showing the bole

Appendix 4. Buttress of the same tree

Appendix 5. Cross section view of wood sample: ***Pometia pinnata*** forma ***tomentosa*** **(N.G.F. 3711) showing both irregular band and vasicentric parenchyma**

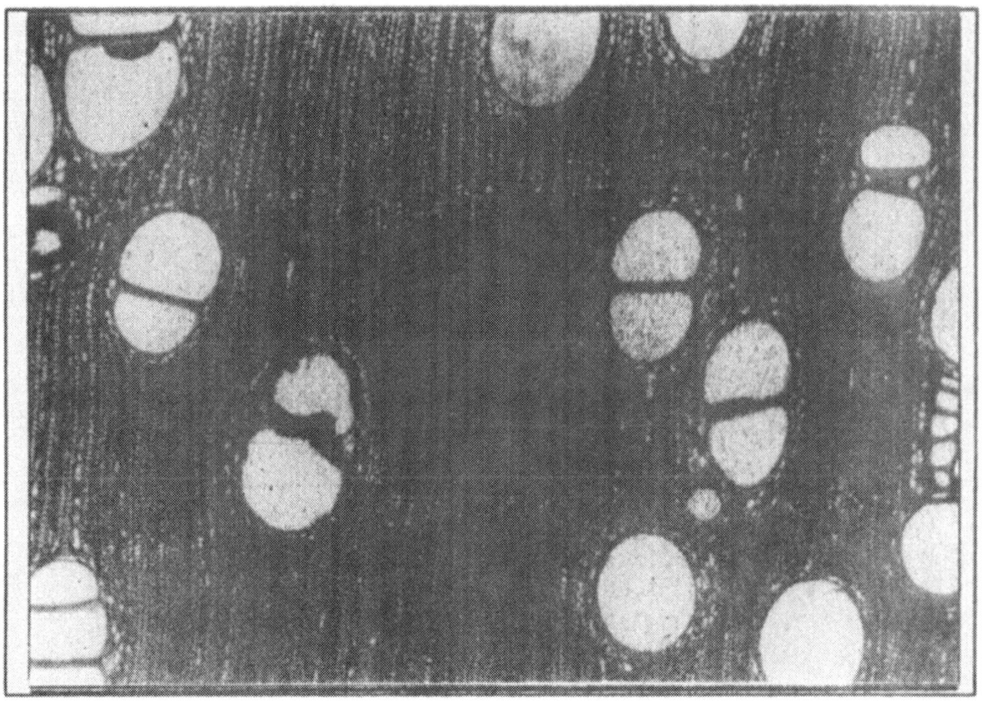

Appendix 6. Cross section view of wood sample: *Pometia pinnata* (N.G.F. 608) **showing only vasicentric parenchyma**

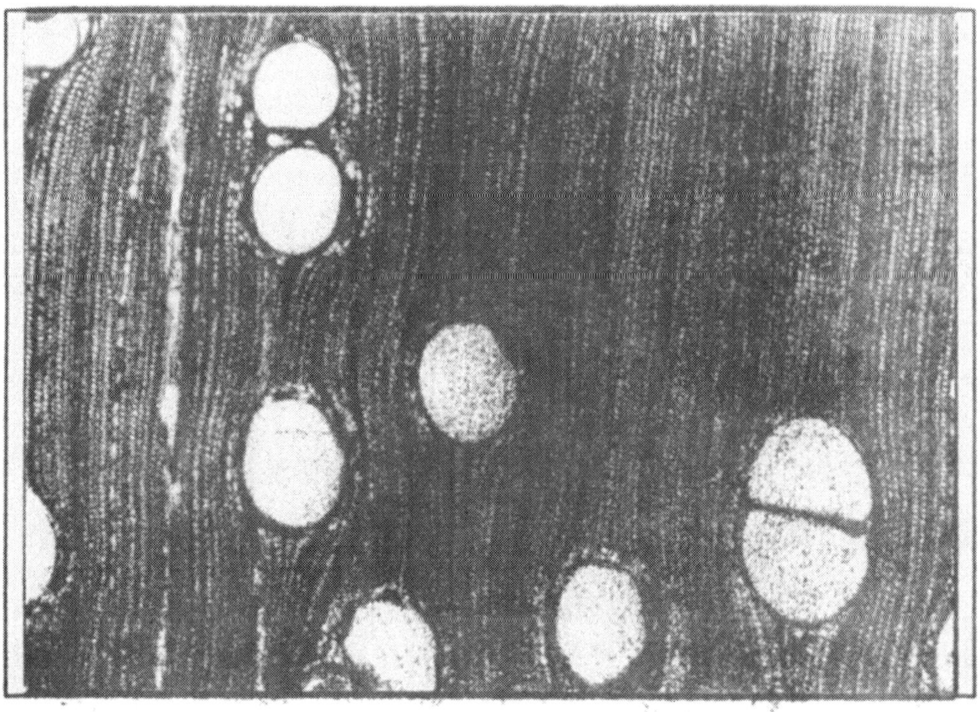

Session III: Ethnomedicine

Proceedings of the Biological Society of New Guinea
R. Höft (ed.)

Preliminary Assessment Of Antibacterial Activity Of Some Traditional Medicinal Plants In Morobe Province, Papua New Guinea

David Timi

Wau Ecology Institute, P.O. Box 77, Wau, Morobe Province, Papua New Guinea

Abstract

As part of the chemical analysis of some traditional medicinal plants used for the treatment of skin diseases, 20 plants locally used for the treatment of wounds and superficial infections were screened for their *in-vitro* antibacterial activity against four bacteria that cause wound infection. The bioassay was carried out by means of the agar overlay technique. Filter paper discs were impregnated with crude water and methanol extracts. Once dried, they were placed over standardized inocula of four indicator bacteria. These included *Bacillus cereus*, *Staphylococcus aureus* (both Gram-positive), *Escherichia coli* and *Pseudomonas aeruginosa* (both Gram-negative). After incubation, the growth inhibition of the bacteria exerted by the plant extracts was measured, demonstrating that 6 of the 20 plants have some noticeable *in-vitro* antibacterial property against one or more of the test organisms.

Introduction

Superficial wound infection is quite common in Papua New Guinea especially in areas where there are poor hygienic conditions and a lack of adequate medical care. Traditionally - and even nowadays - wounds (sores) were treated with natural medicines obtained from plants. The knowledge and practice of these traditional herbal medicines differed among localities. Different plants and plant parts (seeds, roots, barks, leaves, sap, latex, shoots and stem) were used for wound treatment. The nature of treatments varied, too. The most common ways included application of a decoction from a plant part (bark, leaves etc.) to the wounds, application of plant sap or latex, bandaging with heated leaves or with scraped seeds and drinking of water extract. (Woodley 1991).

There is ample scientific evidence that some classes of natural products have antimicrobial activity against organisms causing skin infections. They include alkaloids, terpenoids, organic acids and some phenolic compounds. Not much *in-vitro* examination and chemical and pharmacological studies of the traditional medicinal plants of this country has been undertaken. Thus, there is a need to venture into this area of studies to establish an information centre on scientifically proven medicinal plants possessing medical properties which can serve as an alternative to the costly imported medicines.

METHODS

Ethnobotanical Study

A survey was carried out among traditional medicinal plant users to identify some indigenous medicinal plants commonly used for treatments of sores and fungal infections. From the survey, eleven plants were identified and sampled. Botanical identification of most of these plants were made at Wau Ecology Institute, while the remaining ones were identified at the Papua New Guinea Forest Research Institute in Lae. The other 9 plants reported here were selected from a medicinal plant book (Woodley 1991). No reference collections of the material were made.

Extraction

Five grams of dried (40-50°C), ground samples were extracted with hot water in a waterbath for one hour and filtered. The extraction process was repeated and the extracts were combined and stored in a refrigerator until a day before the tests.

For comparision, 10 grams of fresh tissues of five of the same plants were macerated with methanol in a blender, filtered, evaporated *in vacuo* and stored in a refrigerator. The same methanol extraction procedure was used for the extraction of 8 other plants which were dried at 40-50°C.

Sap and latex were collected a day before the test and stored in a refrigerator.

Preparation of Test Micro-Organisms

With a preflamed inoculating loop the test bacteria *(Bacillus cereus, Staphylococcus aureus, Escherichia coli and Pseudomonas aeruginosa)* were obtained from mother cultures, transferred to sterilized nutrient broth (oxoid) and subcultured overnight. *B. cereus* was incubated at 30°C, *P. aeruginosa* at 25°C. *S. aureus and E. coli* were both incubated at 37°C. The nutrient broth (13 grams) was dissolved in a litre of distilled water and sterilized in an autoclave at 121°C for 15 minutes.

Preparation of Assay Plates and Filter Paper Discs

28 grams of nutrient agar (oxoid) was dissolved in 1 litre of distilled water and sterilized in an autoclave for 15 minutes at 121°C. After sterilization, 15 to 20 ml of nutrient agar were dispensed into plastic petri dishes, having internal diameter of 9 cm. After solidifying, they were placed in an oven at 35°C with the tops slightly open for moisture to evaporate from the dishes. The dishes which were not immediately used were stored in a refrigerator.

6 mm diameter filter paper discs were cut out from a Whatman No.1 filter paper with a paper hole puncher, and were sterilized at a temperature of 120°C in an oven for about 24 hours.

In-vitro *Screening*

0.1 ml of the subcultured suspension was transferred to nutrient agar plates with a sterile 1 ml pipette and streaked evenly with a sterile swab. Filter paper discs were impregnated with plant extracts by touching the extract with the edge of the filter paper allowing the extract to soak the filter paper. After the solvent was dried off, the filter discs were placed in the petri dishes and gently pressed on the agar to ensure complete contact of the discs to the surface of the seeded media. Blank controls were treated with the solvent used to dissolve extract residues. Incubation time for *S. aureus* (37°C), *E. coli* (37°C) and *B. cereus* (30°C) was 24 hours and for *P. aeruginosa* (25°C) 48 hours. After incubation, the inhibition zones were measured with a transparent ruler.

RESULTS

The extent of inhibition of the growth of bacteria by water extracts of 11 plants is shown in Table 1. *Macrococculus pomiferus* and *Flemingia strobilifera* showed obvious bactericidal properties against two of the three test bacteria. *Macrococculus pomiferus* was toxic against *B. cereus* and *E. coli* and *Flemingia strobilifera* was toxic against *E. coli* and *S. aureus*. Five others showed some sign of inhibition around the edges of the filter papers.

Table 1. Antibacterial activity of plant parts after water and methanol extraction and traditional external uses (internal traditional uses are not mentioned here)

Plant	Bacterium:	*E.c.*		*B.c.*		*S.a.*		
	Extraction method:	W.E.	M.E.	W.E.	M.E.	W.E.	M.E.	Traditional treatment
APOCYNACEAE (bark) *Alstonia spectabilis* R. Br		0	-	0	-	0	-	water extract is drunk to treat scabies
BURSERACEAE (bark) *Canarium macadamii* Leenh.		+	-	+	-	0	-	sap is used to treat grille (Tinea imbricata)
CONVOLVULACEAE (bark) *Ipomoea acuminata* (Vahl.) Roem. & Schilter		0	+	0	0	+	++	latex is blown onto fresh cuts
DILLENIACEAE (bark) *Dillenia papuana* L.		+	+	+	0	+	++	sap applied on sores; also used to kill head lice
EUPHOBEACEAE (leaves/stem) ***Euphorbia heterophylla*** L.		+	-	+	-	+	-	latex is applied on sores
LEGUMINOSAE (root) ***Derris elliptica*** (Roxb.) Benth.		0	-	0	-	+	0	infusion from root is used to treat scabies
LEGUMINOSAE (leaves) ***Flemingia strobilifera*** (L.) R. Br. ex Ait. f.		++	-	0	-	++	-	infusion from leaves is applied on sores
MENISPERMACEAE (stem) ***Macrococculus pomiferus*** Becc.		+	-	++	-	0	-	water extract from stem is applied onto ulcers
MORACEAE (bark) *Ficus botryocarpa* Miq.		+	+	+	++	+	++	latex from unripe fruits and bark are applied onto sores
RHAMNACEAE (bark) *Alphitonia incana* (Roxb.) Teys & Binn ex Kurz		0	-	0	-	0	-	not used for skin infection but is used as a liniment
URTICACEAE *(bark)* *Oreocnide rubescens* (Bl.) Miq.		0	+	+	0	0	0	water extract is used to treat scabies
Control		0	0	0	0	0	0	

E.c. = *Escharichia coli*; *B.c.* = *Bacillus cereus*; *S.a.* = *Staphylococcus aureus*; W.E. = water extract; M.E. = methanol extract

- not tested; 0 no growth inhibition; + inhibition diameter (including 6 mm disk) from 6.5-10mm; ++ from 11-20mm; +++ from 21-30mm; ++++ > 30mm.

For comparision, fresh material of five of the same plants was extracted with methanol and concentrated before the test. *Ficus botryocarpa* demonstrated clear bacteria toxic property against all three organisms, followed by *Ipomoea acuminata* and *Dillenia papuana* which were toxic against two test bacteria. Results from the other two plants did not show significant differences with their respective results from the water extracts (Table 1).

In Table 2, the results from methanol extracts of 8 other plants including a reference (procaine penicillin), a blank control and an additional test bacterium *(Pseudomonas aeruginosa)*. Extracts from both bark and leaves of *Graptophyllum pictum* and *Anthocephalus chinensis* showed toxic activity against *P. aeruginosa*, while *Bidens pilosa*, *Erigeron sumatrensis* and *Ocimum basilicum* exhibited toxic property against *Pseudomonas aeruginosa*, *Bacillus cereus and Staphylococcus aureus*. None of the nine plants showed any toxic activity against *Escherichia coli* except the reference compound which also demonstrated bigger inhibition zone against all three other bacteria compared to the plant extracts.

Table 2. Antibacterial effect of methanol extracts from 8 plants and their traditional external uses (internal traditional uses are not mentioned here)

Plants	Plant Part	Bacterium: E.c.	P.a.	S.a.	B.c.	Traditional treatment
ACANTHACEAE						
Graptophyllum pictum (L.) Griff.	leaves	0	+++	0	0	Sap from plant is used
	bark	0	+++	0	0	to treat spear wounds
AMARILLIDACEAE						
Crinum asiaticum L.	leaves	0	0	+	+	Latex from leaf is applied to cuts
COMPOSITAE						
Bidens pilosa L.	leaves	0	++	+	+	Leaf is used as plaster on boils
Erechtites valerianifolia (Wolf) DC.	leaves	0	0	0	0	Juice extract from leaves and stem is applied to sores and scabies
Erigeron sumatrensis Retz.	leaves	0	++++	++	++	Leaves are rubbed onto grille
LABIATAE						
Ocimum basilicum L.	leaves	0	+++	++	+	Infusion of plant is squeezed on sores
Plectranthus scutellarioides (L.) R. Br.	bark	0	0	0	0	Extraction from the plant is used to treat sores and body wounds
RUBIACEAE						
Anthocephalus chinensis (Lamk.) A. Rich. ex Walp	bark	0	+++	0	0	Bark water extract is drunk to treat boils
Procaine Penicillin (reference) *		++	++++	+++	++	
Blank control		0	0	0	0	

E.c. = *Escharichia coli*; *B.c.* = *Bacillus cereus*; *S.a.* = *Staphylococcus aureus*; *P.a.* = *Pseudomonas aeruginosa*; - not tested; 0 no growth inhibition; + inhibition diameter (including 6 mm disk) from 6.5-10mm; ++ from 11-20mm; +++ from 21-30mm; ++++ > 30mm.

Results of extracts from sap and latex are included in Table 3. Filter papers had been directly impregnated with the sap and latex and were placed on seeded agar media. The latex from *Ficus botryocarpa* showed stronger toxic effect against all four test bacteria than water and methanol extracts of the tree bark as shown in Table 1.

Table 3. Growth inhibition by sap and latex of five plants and their traditional medicinal uses

Plant	Plant Part	Bacterium: E.c.	P.a.	S.a.	B.c.	Traditional treatment
Ipomoea acuminata	(latex)	0	0	0	-	see Tab. 1
Dillenia papuana	(sap)	0	0	0	-	see Tab. 1
Ficus botryocarpa	(fruit latex)	++	+++	+++	+++	see Tab. 1
	(bark latex)	0	++	++	+++	see Tab. 1
Piper aduncum	(sap)	+	+	+	-	
Aloe vera (L.) Burm. f.	(juice)	0	0	0	0	Juice from leaf is used to treat athletes foot and sores

E.c. = *Escharichia coli*; *B.c.* = *Bacillus cereus*; *S.a.* = *Staphylococcus aureus*; *P.a.* = *Pseudomonas aeruginosa*; - not tested; 0 no growth inhibition; + inhibition diameter (including 6 mm disk) from 6.5-10mm; ++ from 11-20mm; +++ from 21-30mm; ++++ > 30mm.

Discussion

Pseudomonas aeruginosa seemed more susceptible to the plant extracts than other bacteria. This could be due to its slow growth. Generally, fast-growing organisms have smaller inhibition zones than slow-growing ones (Lorian 1983).

It is also interesting to note that the sap and latex collected from both *Dillenia papuana* and *Ipomoea acuminata*, did not show any inhibatory effect, whereas the methanol extract of their bark and stems showed some effect. The plants are used in Maprik, East Sepik Province in traditional medicine. The transport to Morobe Province may have led to some quantitative changes in the chemical composition and may thus have had influence on the results reported here.

This study supports the results of some similar studies previously carried out on medicinal plants. Caceres *et al.* (1987) reported that 28 of their 89 study plants revealed some *in vitro* inhibition of the tested micro-organisms and Hussain & Deeni (1991) using agar overlay technique from alcohol extracts against 13 indicator orgaisms mentioned, that 30 of the 45 plants screened had antimicrobial properties. Rali (1992) screened 21 plants using bioautographic methods. He reported that all except one had antibacterial activity. In addition to these, Adaay *et al.* (1987) reported on the antimicrobial activity of different fractions of the seed of *Peganum harmala* (Zygophyllaceae), while Janssen *et at.* (1985) and Oguntimein *et al.* (1987) reported on the antimicrobial activity of the essential oils of *Ducrosia anethifolia* (Apiaceae) and *Eupatorium odoratum*, respectively.

Conclusions

Of the 20 plants screened for their *in-vitro* antibacterial activity, 6 exhibited a clear bacterio-toxic property against more than one of the test bacterium. These include *Bidens pilosa (Compositae)*, *Erigeron sumatrensis (Compositae)*, *Ficus botryocarpa (Moraceae)*, *Flemingia strobilifera (Leguminosae)*, *Macrococculus pomiferus (Menispermaceae)*, *Ocimum basilicum (Labiatae)*. Two others, *Graptophyllum pictum* (Acanthaceae) and *Anthocephalus chinensis* (Rubiaceae) were toxic against *Pseudomonas aeruginosa*. The rest did not show any promising antibacterial activity.

Since the study was only a preliminary investigation done without replicates the results may not be reliable. Nevertheless, the bactericidal activity of some of the plants observed from this study, could possibly verify the traditional claims on their healing properties.

ACKNOWLEDGEMENTS

The author wishes to thank the following persons for the contributions they made towards the study. Botanists from Wau Ecology Institute and Papua New Guinea Forest Research Institute for kindly having the plants identified. Olo Gebia and Kairo Alberto, Wau Ecology Institute; Max Kuduk, Paul Katik and Roy Banka, Papua New Guinea Forest Research Institute. The Applied Science Department, Papua New Guinea University of Technology provided laboratory space, facilities for microbiological work and Niamet Kusunan gave technical assistance.

REFERENCES

Adaay, H.M., Rashan, J.L. & Sulayman, D.K. (1989). Antimicrobial Activity of Different Extracts From the Seeds of *Peganum harmala*. *Fitoterapia* **60(4),** 363-365.

Caceres, A., Giron, M.L., Asvarado R.S., & Torres, F.M. (1987). Screening of Antimicrobial Activity of Plants Popularly used in Guatemala for the Treatment of Dermatomucosal Diseases. *J. Ethnopharmacology* **20,** 223-237.

Hussain, H.S.N. & Deeni, Y.Y. (1991). Plants in Kano Ethnomedicine; Screening for Antimicrobial Activity and Alkaloids. *Pharmacognosy* **29,** 51-56.

Ishida M. (unpublished). Preliminary Phytochemical Screening. Wau Ecology Institute; Typescript.

Janssen A.M., Scheffer, .C.J.J., Svendsen, B.A., & Aynehchi, Y. (eds. 1984). Composition and Antimicrobial Activity of the Essential Oil of *Ducrosia anethifolia*. *Essential Oils and Aromatic Plants.* Martinus Nijhoff, Dr W. Junk Publishers, Dordrecht.

Lorian V. (1983). *Antibiotics in Laboratory Medicine.* Williams & Wilkins Company. Baltimore, USA, pp. 24-53.

Inya-Agha, I.S., Oguntimein O.B., Sofowora, A. & Benjamin, V.T. (1987). Phytochemical and Antibacterial Studies on the Essential Oil of *Eupatorium odoratum*. *Int. J. Crude Drug Res.* **25(1),** 49-52.

Rali, T. (1992). Screening of Some Papua New Guinea Medicinal Plants for Antibacterial Activity. *Science in New Guinea* **18(2),** 63-64.

Woodley, E. (ed. 1991). *Medicinal Plants of Papua New Guinea. Part 1: Morobe Province.* Verlag Josef Margraf. Wau Ecology Institute Handbook No. 11.

Proceedings of the Biological Society of New Guinea
R. Höft (ed.)

WATI, A POWERFUL PLANT BEVERAGE - AN ETHNOBOTANICAL CASE OF THE MARIND ANIM, IRIAN JAYA

Markus Warip

University of Cenderawasih, P.O. Box 422, Jayapura 99351, Indonesia

Wati is an endemic plant, cultivated by the Marind Anim. It is honored and given a high status. The plants are looked after more carefully than human babies. Wati is classified according to its colour and the form of the branching. Its taste is an important character. The various plant parts all have their own names.

Wati can be propagated only by persons of high prestige. It is only used in important cultural events and ceremonies such as marriages and funerals. Wati is believed to have contraceptive properties.

SESSION IV: ENTOMOLOGY

Proceedings of the Biological Society of New Guinea
R. Höft (ed.)

PATTERNS OF BEETLE SPECIES DIVERSITY IN *CASTANOPSIS ACUMINATISSIMA* (FAGACEAE) TREES STUDIED WITH CANOPY FOGGING TECHNIQUES IN MID-MONTANE NEW GUINEA RAIN FOREST

Allen Allison, Scott Miller & Allan Samuelson

Dept. of Zoology, Bishop Museum, P. O. Box 1900A, Honolulu HI 99817-0916, USA

Canopy fogging with a pyrethrum-based insecticide was used to study the structure and diversity of insect communities in eight *Castanopsis acuminatissima* (family Fagaceae) trees at elevations of 1200-1400 m in the Wau Valley, Papua New Guinea. Insects were collected in 1 m^2 trays suspended beneath the trees and indentified to morphospecies. The eight trees yielded 3977 individual beetles representing 418 morphospecies in 53 families.

The mean number of beetle morphospecies per tree was 114 (range 82-155). Of these morphospecies, 199 (47.6 percent) were represented by single individuals, and only 62 (14.8 percent) were represented by 10 or more individuals. Approximately a quarter of the species on each tree were restricted to that tree although those species represented only *ca.* 5-10 percent of the individuals. The similarity of the beetle fauna between trees was rather low, ranging from 12 to 31 percent.

This study is part of a much larger canopy fogging project in which 51 trees at study sites of-500 m, 1200-1400 m and 2100-2200m have been fogged. Preliminary analysis of these samples indicates that diversity is similar at the 500 m and 1200-1400 m sites and much higher at the 2100-2200 m site.

Proceedings of the Biological Society of New Guinea
R. Höft (ed.)

NOVEL ANTI-PREDATOR STRATEGIES IN PAPUA NEW GUINEA MOTHS: SPIDER, EVIL FACE, AND PREYING MANTID IMAGES

Larry Orsak

Christensen Research Institute, P.O. Box 305, Madang, Papua New Guinea

Like all moths, the colours, patterns, and resting postures of Papua New Guinea's approximately 10,000 species combine to form an image which deters predators. The main visual predators of moths are birds, lizards and invertebrates (including spiders). Thus, moth restingpostures have largely evolved in response to the pressure these predators produce - the moths that look the least like predator food have had a selective advantage and passed on their genes. Since many of the colour patterns found in moths are found in related species that live outside of New Guinea, some of these patterns have likely evolved over millions of years, predating the island's existence. The step-by-step accumulation of small mutations over this period of time has resulted in some extraordinary moth resting postures. Some preliminary tests on a Pyralis moth [subfamily Nymphuliinae] indicates that its colour pattern will turn out to be the first known butterfly or moth colour pattern that is effective not against vertebrate predators (such as birds), but against invertebrates (specifically. jumping spiders). A noctuid moth (*Ghirdina* sp.) appears to be the first known living thing to have a complete image of a totally different species superimposed upon it. The image seems to be that of a bark-inhabiting preying mantid insect. This probably also represents an example of a novel untested mimicry strategy in which the "model" is not bad-tasting or inedible, but rather, is a lower food reward, compared to the mimic. *Pingasa blanda*, a geometrid moth in Papua New Guinea, seems to follow the same strategy; however, it's model is a bark-inhabiting sparassid (harvestman) spider.

Finally, Papua New Guinean moths display some of the best examples of the "sinister image" described by sociobiologist Robert Trivers. The presumed function of these face-like images is to duplicate a generic predator face that Trivers feels that birds (and humans) have an instinctive fear of.

The study of basic biological phenomena, such as moth defensive strategies, is useful to Papua New Guinea, because the interesting natural history stories that unfold can be used to document just how unique Papua New Guinea's forest wildlife really is. That, in turn, can be used to help attract more tourists to the country... if the forests survive, that is!

Proceedings of the Biological Society of New Guinea
R. Höft (ed.)

Can apparent leaf damage in tropical trees be predicted by herbivore load or host-related variables? A case study in Papua New Guinea

Yves Basset & Robert Höft [†]

Dept of Zoology, Bishop Museum, PO Box 19000A, Honolulu HI 96817-0916, USA
[†] *Wau Ecology Institute, PO Box 77, Wau, Papua New Guinea*

ABSTRACT

Local patterns in leaf damage for ten tree species of different plant families and of different successional status were examined during a one-year study in Papua New Guinea. We investigated whether patterns of interspecific leaf damage on mature leaves could be predicted from either insect-related variables (*i.e.*, species richness, density and biomass of associated herbivores; densities of ants and other enemies) or host-related variables (*i.e.*, simple measurements of plant architecture and biochemistry).

The study was performed on the slopes of Mt. Kaindi, near and within the grounds of the Wau Ecology Institute, Wau. Altitude of collecting ranged from 1,100-2,362 m (summit), but was mostly confined around 1,200-1,400 m. Study trees included: *Elmerrillia tsiampacca* (L.) Dandy (Magnoliaceae), *Cinnamomum cf. culilaban* (L.) Presl (Lauraceae), *Piper plagiophyllum* K. Sch. & Laut. (Piperaceae), *Ficus nodosa* Teys. & Binn. (Moraceae), *Pipturus argenteus* Wedd. (Urticaceae), *Castanopsis acuminatissima* A. DC. (Fagaceae), *Caldcluvia brassii* Hoogl. (Cunioniaceae), *Aleurites moluccana* Willd. (Euphorbiaceae), *Euodia cf. radlkoferiana* Laut. (Rutaceae) and *Cordia dichotoma* Forst. (Boraginaceae).

Leaf-chewing insects feeding externally (*i.e.*, most Lepidoptera larvae, some adults and larvae of the Curculionidae, Chrysomelidae and other beetles, some larvae and adults of the Orthoptera and Phasmatodea) were collected exclusively from the foliage of the study trees by hand collecting, beating and branch clipping. When the foliage of trees could not be sampled from the ground, the single rope technique provided access to the crowns. Sampling effort was identical for each tree species (300 beating samples, about 33 m^2 of foliage clipped and 50 hours of observation).

Insect specialization was assessed from simple feeding trials in the laboratory with the foliage of the study trees. Live insects were stored in plastic vials, at room temperature and in conditions of near-saturated relative humidity. Insects were assigned to three categories according to the results of the feeding tests: (a) "specialists", *i.e.*, insects tested on three or more plant species but which only fed on the plant they were collected from; (b) "generalists", *i.e.*, insects tested on three or more plant species and which fed on two or more plants; and (c) "uncertains", *i.e.*, insects which, because of death or pupation, could not be tested on more than two plant species. An estimate of leaf palatability was obtained by summing all scores recorded during the feeding experiments for a particular tree species and dividing this value by the number of insect species tested on this species.

Since leaf size and leaf weight varied considerably among tree species, three related measurements of apparent leaf damage were considered. First, apparent leaf damage (ALD) was estimated visually and expressed as percentage of leaf area missing. Fifty-five foliage samples were obtained for each tree species, from different individuals and at different periods of the year. For each sample, 15 mature leaves were randomly chosen and scored with the following score classes: 0: 0 percent, 1: 1 percent, 2: 2-5 percent, 3: 5-10 percent, 4: 10-20 percent, 5: 20-50 percent, 6: 50-80 percent and 7: 80-100 percent. The average ALD of the sample was calculated as the average percentage damage per leaf, obtained by summing the mid-point value of corresponding class scores and dividing this sum by the number of leaves.

Second, a crude estimate of area removed per leaf (ARL) was calculated as follows:

$$ARL\ (cm^2) = ALD/100 \times \text{Mean specific leaf size (double-sided area, } cm^2)$$

Third, a similar estimate of biomass removed per leaf (BRL) was calculated as:

$$BRL\ (g\ DW) = ALD/100 \times \text{Mean specific leaf size } (cm^2) \times \text{Mean specific leaf weight } (g\ DW \times cm^{-2})$$

Several plant-related attributes, which could potentially influence leaf damage and which could be easily determined in the field or from the literature, were considered for each study tree. The variables were classified into the following categories: (1) taxonomic isolation of the host; (2) variables related to leaf production; (3) variables related to growth strategy: successional status (pioneer or persistent species), deciduousness, altitudinal range, and local abundance; (4) variables related to architecture: mean tree height, mean leaf/leaflet area, specific weight of mature leaves and leaf pubescence; (5) variables related to simple aspects of plant chemistry: mean water content of mature leaves, total foliar nitrogen and potassium, and leaf palatability

When data were pooled for the ten tree species studied, mean ALD amounted to 12.0 ± 0.4 percent per sample, mean ARL to 16.3 ± 0.6 cm^2 per leaf and mean BRL to 0.16 ± 0.01 g DW per leaf. Leaf damage varied greatly among species, from 3.8 to 19.7 percent, ALD, from 5.7 to 28.9 cm^2 ARL, and from 0.05 to 0.50 g DW BRL. Interspecific differences in leaf damage varied greatly and depending on how measurement was reported, species showing low and high levels of damage were different. When the frequency distribution of leaf damage was examined for each tree species, most distributions were different from each other. All distributions were positively skewed, but to various degrees. Interestingly, there was a negative correlation between ALD and skewness of distribution. Damage expressed as percentage of leaf area lost was often involved in the few significant correlations with insect- and host-related variables (see below) and seemed therefore to be the most appropriate variable to use.

Leaf damage was difficult to predict from insect-related variables. Species richness and specialization of associated insect herbivores did not account for the levels of leaf damage observed. As a result, trees with species-rich herbivore faunas or with high herbivore loads were not always likely to experience high damage. Patterns in leaf damage appear more likely to depend on the feeding behaviour of a few dominant herbivore species and at which time the damage is caused than on herbivore pressure expressed as species richness. Analyses are currently under way to assess the influence of herbivore density and biomass on leaf damage, as well as that of the density of insect enemies.

Leaf damage was also difficult to predict from host-related variables (tree architecture, water and nitrogen content of leaves, etc.). Contrary to expectations (see Coley 1983), leaf palatability did not clearly influence leaf damage, nor did successional status or leaf production patterns. The strongest correlations that we could establish in our data set were between ALD and the skewness of damage distribution (a negative correlation), and between both ALD and skewness and specific leaf area (negative and positive correlations, respectively): trees with large leaves were less likely to experience high damage levels overall and on these trees the distribution of damage was highly positively skewed. We believe that this may result from the foraging and feeding behaviour of insect herbivores which may be particularly conspicuous to enemies on trees with large leaves. In this instance, it seems to be advantageous for the herbivore to proceed by isolated, distant, feeding bouts in which a high quantity of foliage is ingested, in order to spread leaf damage and to escape from visually-oriented predators, such as birds.

Our findings can be discussed with reference to the apparent paradox observed in tropical rain forests (Elton 1973): although insect herbivores are scarce there, leaf damage is relatively high. However, at the moment, the interpretation of our data is not definitive, because certain important insect variables, such as insect biomass, have not been determined for all tree species yet. With the data in hand, we suggest that leaf damage on a particular tree species is likely to depend on the feeding behaviour of a few dominant leaf-chewing species. This, in turn, may complicate the interpretation of results obtained in studies of herbivory seeking community-level patterns.

REFERENCES

Coley, P.D. (1983). Herbivory and defensive characteristics of tree species in a lowland tropical forest. *Ecological Monographs*, **53**, 209-233.

Elton, C.S. (1973). The structure of invertebrate populations inside neotropical rain forest. *Journal of Animal Ecology*, **42**, 55-104.

Proceedings of the Biological Society of New Guinea
R. Höft (ed.)

NOTE ON STUDIES OF THE BREEDING BEHAVIOUR OF TARO BEETLE *PAPUANA* SP. IN PAPUA NEW GUINEA

Sim A. Sar & Benjamin Niangu

Bubia Agricultural Research Centre, P.O. Box 1639, Lae, Papua New Guinea

INTRODUCTION

The Taro beetle, *Papuana* sp. [Coleoptera: Scarabaeidea] is the most serious pest of taro *Colocasia esculenta*. The genus *Papuana* has about 20 species, of which most are native to the Papua New Guineas sub-region (Waterhouse & Norris 1987). The most important species in Papua New Guinea are *P. woodlarkiana*, *P. biroi*, *P. huebneri*, *P. szentivani* and *P. uninodis*, The adult stage feeds on the taro corm while the larva is found in the soil.

MATERIALS AND METHODS

Samples of the different life stages of the beetle were collected by digging an area of $0.5m^2$ under different grass species in each sampling site and recording the numbers of eggs, larvae, pupae and adults. The sites selected were in the major taro growing areas and included Labu-Butu, Wapi outside, West Taraka-Lae and Ticleling in the Bukawa area. Grasses were collected for identification while soil samples were collected for soil analysis.

RESULTS

The results of the survey show that eggs are deposited under certain grass species which include *Impereta cylindrica*, *Saccharum robustum*, *Ophiuros tongcaling*, *Paspalum paniculatum*, *Sorghum verticilliflorum* and *Pennisetum purpureum* (Fig. 1). Higher numbers of larvae were collected from *P. purpureum*, *S. verticilliflorum* and *P. paniculatum*. Most of the larvae were collected in free-draining soils.

DISCUSSION

The results of this survey has confirmed the breeding habitat of taro beetle. The adults feed on taro before dispersing to favourable breeding sites. Immature stages are rarely found in taro gardens. Although taro beetle larvae have been collected under these grass species and other similar habitats (Perry 1977), other factors are responsible for the distribution of larvae in the soil. Soil structure and soil organic matter, perhaps, determine larval density in various taro beetle breeding habitats. In support of this view it has been found that taro beetle larvae are rarely found in heavy clay soils nor in light sandy soils. Larvae are very sensitive to changes in the moisture level. Light to medium free draining soils are probably preferred. Furthermore, taro beetle larvae have been collected under decaying kitchen refuse, chicken manure and heaps of coconut husks.

Proceedings of the Biological Society of New Guinea
R. Höft (ed.)

THE LIFE CYCLE OF *ORNITHOPTERA PRIAMUS POSEIDON* DOUBLEDAY [LEPIDOPTERA: PAPILIONIDAE] AT THE FULICA BUTTERFLY FARM IN MANOKWARI

S. Joris, J.P. Nerokouw, R. Angrianto & Alexander Yaku

Faperta UNCEN Manokwari, P.O. Box 23, Manokwari 98314, Indonesia

ABSTRACT

The life cycle of the bird wing butterfly, *Ornithoptera priamus poseidon* Doubleday was studied at the Fulica Butterfly Farm in Manokwari. The observations were carried out over four months.

The results show that the life cycle of adult males and females of the *O. priamus poseidon* from egg to imago can take place in 76 days and 93 days, respectively.

Eggs can hatch in 6 days and emerging larvae live for 26 days. During this period, larvae undergo four moultings. During the first instar to third instar, larvae feed on leaves of *Aristolochia*, and during the fourth and fifth instar larvae feed on twigs and branches of the plant as additional food.

The prepupa stage occurs in three days, and at that time larvae move to the broader leaves in order to rest and transform to the pupa. The pupa stage takes place in 29 days and the adult male and female that emerge can live for 12 days and 29 days, respectively.

Proceedings of the Biological Society of New Guinea
R. Höft (ed.)

MAIN INSECT PESTS OF MAIZE IN THE PAPUA NEW GUINEA LOWLAND; THEIR PREDATORS AND PARASITOIDS: AN OVERVIEW

Jan Breithaupt

Bubia Agriculture Research Centre P.O.Box 1639, Lae, Papua New Guinea

ABSTRACT

The most serious insect pests of maize in the Papua New Guinea lowlands are the Asian Corn Stem Borer, *Ostrinia furnacalis* Guenée [Lepidoptera: Pyralidae], followed by the Earworm, *Helicoverpa armigera* Hübner [Lepidoptera: Noctuidae] and three armyworm species, *Spodoptera exempta* Walker, *S. littura* L. and *S. mauritia* Baisdural [Lepidoptera: Noctuidae].

A jumping spider of the Oxipidae family plays an important role in reducing L1-, L2-, L3- and L4-larvae populations in the field. Its average number of prey fed per day rises up to 0.8 larvae. Hence, predation on larvae reaches up to 12 percent in the field.

The local egg parasitoid *Trichogramma* nr. *plasseyensis* Nagaraja [Hymenoptera: Trichogrammatidae] is the most successful parasitoid of *O. furnacalis* and *H. armigera*, reaching parasitization rates between 70 and 100 percent during both, wet and dry season.

INTRODUCTION

The objectives of the biological control project at Bubia Agricultural Research Centre are to identify the major pests of corn in Papua New Guinea and to review the natural enemies of the major pests of corn including the evaluation of parasitoids connected with these insect pests.

All attempts to control the major insect pests in corn chemically were rather ineffective and too expensive for smallholders (Young 1992).

Therefore more data on interaction between natural enemies and hosts are needed if consistent control of corn pests is to be achieved (Napompeth 1990; Young 1992; Kumar 1992).

The presented paper gives an overview of the major insect pests, their predators and parasitoids and can be used as the basis for a classical biological control concept of the major pests in corn n the Markham Valley.

MATERIALS AND METHODS

Bubia Agricultural Research Centre is based 15 km outside Lae in the Markham Valley, Morobe Province, 146°E, 6.51°S at an altitude of 16 m a.s.l. The centre receives an annual average rainfall of 2870 mm from both the southeast and northwest monsoons.

The other three field sites are further up the Markham Valley in Erap, Mutzing and Ramu, where the annual rainfall is less.

The maize varieties "Suwan 1" and "Metro" were sown during 1992/93 growing season.

During a total of 15 field trials, insects were collected from the field. The fields of 0.5 ha each were monitored during wet (May-July) and dry season (September-November) throughout one year (June 1992 - June 1993). Most insects were collected from the control plots, where no treatments were applied and plants were allowed to become infested with pests.

The pest status as well as predation and parasitization have been confirmed in various laboratory studies. For insects where the pest status is not known a confirmation test was undertaken: Fresh parts of the maize plant were placed together with one insect pest-candidate each per pill box. For each box, a different part of the maize plant was supplied and then it was observed if any parts were eaten by the insect.

For all tests a sufficient moisture in the pill boxes was ensured.

To confirm predation, predators were collected from the field and one each kept in pill boxes. A drop of water in each box was placed to prevent dehydration of the arthropod. After 24 hours, larvae of different pest species and of different instars were introduced. After confirming predation, equal numbers of each prey were introduced to another set of predators starved for 24 hours to determine prey preference. The same method was used to determine average number of prey fed per day.

Predators and pests were handled by using a fine camel hair brush and pill boxes were used in the field to catch predators.

To detect all parasitoids of the main insect pests of maize, each 100, larvae, pupae and egg (masses) of the pest concerned were collected randomly throughout the year and all the maize growth stages.

Eggs were kept in gelatine capsules and the hatching of larvae or egg parasitoid was monitored daily. Collected pupae and larvae were provided with a fresh plant part of maize and held singly in 10 x 2.5 cm glass vials.

All collected insects were kept in the Bubia Insectery at 25 °C, well protected from ants.

Appendix 1 shows a list of addresses of persons who have been consulted for insect identification.

RESULTS

The major pests of corn, grown in Markham Valley were identified as follows:

Table 1. Most important Pests and their frequency (%) during 1992/1993 maize growing season (DAS) in the Papua New Guinea lowland

Insect Pest	Frequency of pest during growth stages (in percent) and Number of larvae per plant (in brackets)						
	30	40	50	60	70	80	90 DAS
1.*Ostrinia furnacalis* Guenée	2	25	60	75	70	65	70
[Lepidoptera: Pyralidae]	(0.1)	(0.1)	(0.4)	(4.3)	(6.0)	(5.8)	(0.4)
2.*Helicoverpa armigera* Hübner	3	20	10	15	20	20	10
[Lepidoptera: Noctuidae]	(0.1)	(0.1)	(0.1)	(0.9)	(1.7)	(1.8)	(0.1)
3.*Spodoptera* spp. }	55	40	15	2	2	-	-
[Lepidoptera: Noctuidae]	(2.2)	(0.2)	(0.1)	(0.1)	(0.2)		
S.mauritia Boisduval	30	20					
S.exempta Walker	15	10	5	1	1		
S.litura Fabricius	10	10	10	1	1		
4.*Mythimna* spp. }	30	-	-	-	-	-	-
[Lepidoptera: Noctuidae]	(1.2)						
M.separata Walker	25						
M.loreyi Dup.	5						

5.*Agrotis ipsilon* Hufnagel	6	-	-	-	-	-	-
[Lepidoptera: Noctuidae]	(0.2)						
6.Orthoptera gen. }	4	5	5	5	2	10	10
Locusta spp.}	(0.2)	(0.0)	(0.0)	(0.3)	(0.2)	(0.9)	(0.1)
[Orthoptera: Acrididae]	-	-	1	2	1	4	4
Locusta migratoria L.							
Locusta solitarious							
Locusta austracurs							
Oryetes centaurus							
Patanga succincta L.							
Nephotettix nigropictus Shih.	-	3	4	3	1	6	6
Gryllotalpa africana Beauv.	4	2	-	-	-	-	-
[Orthoptera: Gryllotalpidae]							
7.Aphididae sp.;	-	5	10	-	1	2	5
e.g. *Rhopalosiphum maidis* Fitch	(0.0)	(0.1)	(0.1)	(0.2)	(0.0)		
[Hemiptera: Aphididae]							
8.*Protaetia fusca* Herbst	-	5	-	3	5	3	5
[Coleoptera: Cetoniidae]		(0.0)		(0.2)	(0.4)	(0.3)	(0.0)

Various other species, associated with corn, have been identified. Table 2 gives an overview.

Table 2. Overview of identified pests in corn in the Markham Valley

Insect	Behaviour
Nezara viridula L.[Hemiptera: Pentatomidae] (green vegetable bug and its nymps)	Feeding on young shoots
Riptortus rubronotatus Blote. [Hemiptera: Coreidae]	Feeding on young leaves, silk
Peregrinus maidis Ashmead [Hem.: Delphacidae]	Feeding on leaves
Dysdercus sidae Montrouzier [Hemiptera: Pyrrhocoridae]	Feeding on leaves
Dindyrus sp. [Hemiptera: Pyrrhocoridae]	
? [Hemiptera: Flatidae]	Stem feeding
? [Hemiptera: Hydrometridae]	Feeding on leaves
? [Hemiptera: Miridae]	Feeding on flower, silk
Amata hyalota (Meyr.) [Lepidoptera:Amatidae]	Larvae attacking leaves
Euproetis sp. [Lep.: Lymantridae]	Last corn stages, larvae feeding on leaves
Chrysodeixis chalcites (Esper) L. [Lep.: Noctuidae](Corn semilooper)	Larvae feeding esp. on the silk
Scirpophaga excerptalis (Walker) [Lep.: Pyralidae]	Larvae feeding on leaves
Trichoplusia ni Hb. [Lepidoptera: Noctuidae]	Feeding on leaves
Pseudaletia separata (Walker) [Lepidoptera: Noctuidae]	Feeding on cobs
Dacus papuensis Malloch [Diptera: Tephritidae]	Feeding on leaves
Gesonula mundata (Walker)	Feeding on leaves

[Orthoptera: Eumastacidae]	
? [Orthoptera: Gryllacrididae] (field cochroach)	Feeding on leaves, esp.on corn basis
Oribius inimicus Marshall [Coleoptera: Curculionidae]	Feeding on leaves
Rhabdocelus obscurus Boisd. [Coleoptera: Curculionidae](weevil borer)	First growth stage: feeding esp. on roots
? [Coleoptera: Curculionidae]	Feeding on cob
Casnonidia sp. [Coleoptera: Cerambycidae]	Feeding on young leaves, attacking corn basis/on
In the soil	
Aulacophora abdominalis Fabricius [Coleoptera: Chrysomelidae] (pumpkin beetle, secondary pest)	Feeding on young leaves and roots
Aulacophora similis Oliver [Coleoptera: Chrysomelidae]	
Chaetocnema basalis Baly [Coleoptera: Chrysomelidae]	Feeding on leaves
Monolepta nigroapicata Bryant [Coleoptera: Chrysomelidae]	Feeding on tassels
Arsipoda tenimberensis Jacoby [Coleoptera: Chrysomelidae]	Feeding on leaves
Papuana biroi Endrodi [Coleoptera: Scarabaeidae]	In soil: larvae and adult feeding on roots
Lepidiota reuleauxi Brenske/ Arrow [Coleoptera: Melolonthidae] (white grub)	Feeding on roots and young leaves
Epilachna sp. [Col.: Coccinellidae] [e.g.:*E. signatipennis* Bd.; *E. haemorrhea* Bd. *E. doryca* Bd.; *E. guttatopustulata* Fab.]	Feeding on leaves
Spilocaria bissellata Mulsant [Col.: Coccinellidae]	
Cicada gen. [Homoptera: Cicadidae] *Baeturia papuensis* De Boer (green cicada)	In soil: larvae feeding on root system
Baeturia famulus Dist.	
Diaspine scale [Homoptera: Diaspididae]	Laying eggs on leaves
Lophops sp. [Homoptera: Lophopidae]	Feeding on silk

The Asian Corn Stem Borer, *O.furnacalis* is the major insect pest of maize in the Papua New Guinea lowland; causing high production losses of 3000 tonnes per year.

The population density peaked with six larvae per plant at 70 days after seeding (DAS)/ (Table 1).

H.armigera is the second serious insect pest in maize grown in the Papua New Guinea lowland while the population dynamic of Spodoptera spe- cies showed to be tolerable and appears only during the first maize growth stages.

Natural populations of predators play an important role in controlling Asian Corn Borer- and Earworm populations in the field.

Tables 3 and 4 provide information and an overview of the most important predators in lowland fields and their frequency (%).

Table 3. Most important Predators and their frequency (%) during the 1992/ 93 maize growing season in the lowland of Papua New Guinea (average data of five lowland sites):

Predator	Frequency (%)	Ocurrence/growth stage
1. *Oxyopes* sp. [Araneida: Oxipidae]	85	3 - 9
2. *Solenopsis geminata* Fabricius *Polyrhachis* sp. [Hym.: Formicidae]	5	3 - 9
3. Coccinelidae [Col.: Coccinelidae]	3	6 - 9
4. *Orius* sp.[Hemiptera: Anthocoridae] *Menida formosa* Westw[Pentatomidae]	3	4 - 9
5. *Chelisoches morio* Fabricius *Proreus similans* Stal [Dermaptera: Chelisochidae] *Euboriella stali* D. [Labiduridae]	2	5 - 9
6. Mantidae gen. [Mantodea: Mantidae]	1	4 - 9

The number of predators per plant is slowly increasing within the maize growth stages with a maximum of around two predators per plant at 80 days after seeding.

Spider species, especially of the Oxipidae family showed to be the most important predator of various larvae in the field.

Oxyopes sp. preyed on L1-, L2-, L3- and L4-larvae of *O. furnacalis* and *H. armigera* with an efficiency of 0.55-0.73 killed larvae per day and spider at 80 days after seeding. Hence, with an estimated number of 9,300 spiders per hectar, up to 6,800 larvae per day and hectar can be killed, which is equivalent to a predation rate of up to 11.5 percent.

The role of general predators is definitely important to complement the impact provided by the parasitoids. Various other predator species, associated with maize, have been collected from the field and were identified (Table 4): List of identified predators of corn in the Markham Valley.

Table 4. Overview of identified predators in the Markham Valley

	Insect Behaviour
Coleoptera: *Coccinellidae*: spp. *Menochilus sex-maculatus* Fabricius spp. *Micraspis crocea* Mulsant spp. *Micraspis discolor* F. spp. *Leptothea osikii* Wsc. spp. *Coelophora inaequalis* (F.) spp. *Coelophora mulsanti* Montrz. spp. *Coccinella repanda* Thunberg spp. *Coccinella arcuata* (Fab.) spp. *Hormonia octomaculata* (Fab.) spp. *Cryptolaemus sinestria* Booth & Pope spp. *Cryptolaemus wallacei* Crotch spp. *Chilocorus nasicornis* Kors. spp. *Leis papuana* Crotch	Preying on aphids; Predator of egg masses
? [Coleoptera: Coccinellidae]	In silk of corn, esp.predator of *Helicoverpa* eggs
Cassida diomma Baisdural	Feeding on lice, [Coleoptera: Cassidinae] (but feeding on leaves too)
Metioche albovittata [Orthoptera: Gryllidae]	Predator of *Agrotis* sp. eggs

Proreus similans Stal [Dermaptera: Helisochidae]	Predators of eggs, larvae, pupae
Chelisoches morio Fabricius [Dermaptera: Helisochidae]	
Euborellia stali Dohrn [Dermaptera: Labiduridae]	
Tectoris sp. and its nymphs [Hemiptera: ?]	Predator of eggs, larvae (esp. of *Helicoverpa* sp.)
Menida formosa Westwood [Hemiptera: Pentatomidae]	Predator of insect eggs and larvae
Orius sp. [Hemiptera: Anthocoridae]	
Leptoglossus australis Fabricius [Hemiptera: Coreidae]	
Solenopsis geminata Fabricius [Hymenoptera: Formicidae]	Predator of eggs, pupae and larvae
Polyrhachis sp. [Hymenoptera: Formicidae]	
Pheidologeton sp. [Hymenoptera: Mymmicinae]	
Vespa affinis [Hymenoptera: ?]	
Araneida gen. [Araneida: Argiopidae]	Preying on many different insect species
? [Araneida: Salticidae]	
Oxyopes sp. [Araneida: Oxipidae]	Most important predator
Pantala flavescens [Ordonata: Libellulidae]	Predator of many insect species
Mantidae sp. [Mantodea: Mantidae]	Prey on many different insect species
Chrysopa sp. [Neuroptera: Chrysopidae]	Larvae feed on thrips and other species
Bird	**Behaviour**
Milwus migrans (black kite) (hook type)	Eats locusts, armyworm larvae, etc.
Frog	**Behaviour**
Litoria angiana [Rana: Hylidae]	Preying on various adult moths, etc.

Table 5 gives an overview of identified egg-, pupal- and larval parasitoids of the major insect pests in maize, the Asian Corn Stem Borer and the Earworm.

Egg-, pupal- and larval parasitoids of *O. furnacalis* were active throughout the year. The parasitization rates remained stable for all parasitoids with no significant differences between the wet- and the dry season.

The local egg parasitoid *Trichogramma* nr. *plasseyensis* Nagaraja is the most important natural enemy of ACB and Earworm in the Markham Valley, reaching parasitization rates between 70 and 100 percent during both, wet and dry season.

Pupal- and larval parasitization was generally lower; parasitism rates above five percent were reached by *Euplectrus xanthosephalus* Ferriere, a polyphagous ectoparasitoid of *O. furnacalis* larvae and by *Cotesia marginiventris* Cresson. Besides, the pupal parasitoid

Brachymeria lasus Walker is of interest to reduce the borer population in the field.

During the wet season the nematode *Psammomeris canterburis* plays an important role to control the third and fourth instar larvae of the host. The same applies to a *Nuclear Polyhedrosis Virus* (NPV), especial- ly during the dry season. Its larval parasitization peaked at 22 percent and was even higher in the highland sites.

The parasitization rates of all other parasitoids mentioned in Table 5 were below 4 percent and therefore did not play an important role in controlling the Asian Corn Stem Borer as well as other major insect pests in maize grown in the lowland.

Table 5. Overview of identified egg, pupal and larval parasitoids of *Ostrinia furnacalis* Guenée [Lepidoptera: Pyralidae]

Insect	Behaviour
Egg parasitoids	
Trichogramma plasseyensis Nagaraja [Hymenoptera: Trichogrammatidae]	Reared from eggmasses; found in corn and wild canes, esp. *Rottboellia cochinchinensis* Lour. Cl. Attacks *O. furnacalis*, *Sesamia* sp. and *Chilo terrenellus* P., *C. infuscatellus* S. egg masses as well as *H.armigera* eggs
Trichogramma chilotraeae Nagaraja & Nagarkatti [Hymenoptera: Trichogrammatidae]	
Trichogramma australicum Girault [Hymenoptera: Trichogrammatidae]	
? [Hymenoptera: Encyrtidae]	Polyphagous egg parasitoid specific for aphids
Telenomus dignus Gahan [Hymenoptera: Scelionidae]	Reared from *C.terrenellus* and *O.furnacalis* egg masses
Telenomus remus Nixon [Hymenoptera: Scelionidae]	Esp. egg parasitoid of *Spodoptera* sp.
Pupal and larval parasitoids	
Carcelia senometopia R.	Pupal parasitoid of *Ostrinia furnacalis*, *Helicoverpa* and *Sesamia* sp. Egg and larval parasitoid of *Agrotis ipsilon*.
Carcelia carcelia R. [Diptera: Tachinidae, Carceliini]	
Winthemia sp.	Parasites in larval stage. Emerges in pupal stage
Sarcophagidae gen. [Diptera: Sarcophagidae]	Parasites L2/ L3
Cloropidae sp. [Diptera: Sarcophagidae]	On larvae & pupae (esp. L2/ L3)
Cotesia marginiventris Cresson [Hymenoptera: Braconidae]	Larvae parasitoid of *Ostrinia*- and *Sesamia*
Cotesia flavipes Cameron	Larvae parasite esp. of *Sesamia grisescens*
? [Hymenoptera: Braconidae]	Polyphagous pupal parasitoid, esp. on *Nymphalidae*
Gryon sp. [Hymenoptera: Scelionidae]	Pupae parasitoid
Podagrion sp. [Hymenoptera: Torymidae]	Polyphagous parasitoid. Egg parasitoid of *Mantidae*. Pupae parasitoid of *O.furnacalis* (hyperparasite)
Euplectrus sp. [Hymenoptera: Eulophidae] *Euplectrus xanthosephalus* Girault	Polyphagous larval parasitoid
Brachymeria lasus (Walker) [Hymenoptera: Chalcididae]	Polyphagous pupal parasite, esp. on *Nymphalidae*
Brachymeria phya (Walker)	Esp. on *Noctuidae*

Nematodes:	
Psammomeris canterburis (postparasitic juvenile) [Mermithidae/Meloidogyne]	Larval parasitoid of 3+4 instar larvae of the host
Virus:	
NPV (Nuclear Polyhedrosis Virus) [Granulosis: Baculoviridae]	Esp. on *Helicoverpa armigera* larvae
Saphrophagous:	
? [Coleoptera: Staphylinidae] (rove beetles)	Saprophagous esp. on pupae
Phorid [Diptera: Phoridae]	Saprophagous on pupae
? [Diptera: Tipulidae]	Larvae feeding inside stem and on pupae

DISCUSSION

An IPM strategy should be developed to control the major insect pests of maize, the Asian Corn Stem Borer, *O. furnacalis*, and the Earworm, *H.armigera*. Their damage to maize is causing high yield losses.

The role of general predators such as *Oxyopes* sp. should be considered as an important part of the biological control of the major insect pests in maize.

ACB-population can be very well controlled, just by releasing a sufficient number (150,000-300,000 wasps/ha) of the very effective local parasitic wasp *T.plasseyensis* at the right time around 30-40 DAS before the appearance of the first ACB egg masses in the field.

More concentration should be directed to determine the level of egg parasitization and predation to control the *O. furnacalis* and *H. armigera* population in the field.

ACKNOWLEDGEMENTS

I have to thank Mr. Benjamin Niangu and Ronnie Kokorime for their technical assistance. The SPC/GTZ and Bubia Agricultural Research Centre/D.A.L. finance this biological control project.

REFERENCES

Kumar R. (1992). Microbes and Microbials: The Resurgence of an era in Biological Control. *Discovery and Innovation* **4(2),** 4-12.

Napompeth B. (1990). Use of Natural Enemies to control agricultural pests in Thailand. Extension Bull. No. 303, 22pp.; NBCRC, Kasetsart University, Bangkok.

Young G.R. (1992). Some Host Plants, and an Egg Parasite, of the Maize Stem Borer, *Ostrinia furnacalis* (Lepidoptera: Pyralidae) in New Guinea. *Science in New Guinea* **18 (1).**

APPENDIX

Addresses of Consultants for Insect Identification.

International Institute of Entomology,
Identification Service
c/o British Museum (Natural History)
Cromwell Road, London; S.W.J. / U.K.
c/o Dr. John La Salle (Chalcididae sp.)
Andy Polaszek (Spider sp.)

Queensland Museum
P.O.Box 300 Brisbane
c/o Dr. Geoff Montemeith (Senior Currator/
Lower Invertebrates)
Qld. / Australia

Dr. H. Nagaraja (Parasitoids)
former: Ramu Sugar LTD./ Gusap,
P.O.Box 2183 Lae/ P.N.G.
India : No.:233, 8th Main Road; 16th Cross
Malleswaram
Bangalore- 560 055/ India

Dr. Banpot Napompeth (Parasitoids)
NBCRC; Kasetsart University
P.O.Box 9-52; Bangkok 10900/ Thailand

Karl Koenig (Trichogramma sp.)
PGBPPP; Bureau of Plant Industry
692 San Andres St., Malate, Metro Manila
1004
P.O.Box 1021 MCPO Makati/ Philippines

Dr. Peter Ooi (Fungi, etc.)
International Institute of Biological Control/
Malaysian Station
Level 3, Block E, Kompleks Pejabat
Damansara,
Jalan Dungun, Damansara Heights,
50490 Kuala Lumpur/ Malaysia

Dr. Harry Evans (Fungus)
IIBC.
Silwood Park
Ascot, Berks. SL577A/ U.K.

Mr. Bert Barrion (Spider)
IRRI, P.O.Box 955/ Philippines

Dr. Robbin Bedding (Nematode)
Division of Entomology
CSIRO (Black Mountain Laboratories)
GPO. Box 1700
Canberra, ACT 2601/ Australia

Dr. G. Poinar (Nematode)
Dept. of Entomological Science
University of California,
Berkeley - California Asoc. Prof. Jim
Menzies (Ants sp.)
Biology Dept.
Papua New Guinea University; Port
Moresby/ P.N.G.

Oxford Institute of Virology
Mansfield Road
Oxford OX1 3SR/ U.K.

Proceedings of the Biological Society of New Guinea
R. Höft (ed.)

THE USE OF PLANT-DERIVED PESTICIDES IN INTEGRATED PEST MANAGEMENT IN PAPUA NEW GUINEA

Peter Förster

Entomology Section, Lowland Agricultural Experimental Station Keravat, P.O. Box Kerevat, East New Britain Province, Papua New Guinea

INTRODUCTION

Intergrated Pest Management (IPM) is a combination of methods which makes use of all economically, ecologically and toxicologically justifiable means to keep pests below the economic injury level. Emphasis is placed on the use of natural limiting factors.

Since 1973, the FAO (1974) recommends IPM concepts for pest control. This is because it has become obvious that the unlimited use of chemical pesticides is leading to severe environmental impacts and may result in resistance to pest (Brader 1989). IPM intergrates different methods (cultural techniques, physical, biological, bio-technical and chemical control) to control pests and prevent them from damaging agricultural produce.

PRESENT SITUATION OF IPM IN PAPUA NEW GUINEA

So far the use of chemical control in Papua New Guinea is limited - in contrast to many other similar countries elsewhere in the tropics (Africa, Asia). There are several reasons for this fact: the restrictive policy, the small internal market and the bad infrastructure of the country result in high prices for synthetical pesticides. The frame conditions are changing and due to the rapid population growth and the consequently more intensive use of land the importance of pest control for saving the yields will increase (Cook *et al.* 1989). The most simple method which pretends to show fast succes in controlling pests is the use of synthetic pesticides. However, the disadvantages of their intensive use are well known [health hazards; contamination of the resourses (Dinham 1993); impact of natural antagonist (Hassan *et al.* 1987, Förster 1991); resistance of pest (Smith & Pimental 1978)]. Keeping this in mind, the farmers in Papua New Guinea have the chance not to repeat the mistakes which have been made in other countries but to learn from these. As long as pesticides in Papua New Guinea are expensive farmers can be trained in IPM methods which offers alternatives to the costly imported products.

THE TRADITIONAL SYSTEMS

In the traditional smallholder gardens, IPM methods have been developed empirically. The systems were balanced by the use of 10-15 years of fallow. Nowadays, in densly populated areas the time of fallow is reduced to 3-4 years (Cook *et al.* 1989) and the systems are changing by the introduction of new crops, partly as cash crops (cabbages, spices, balsa). Moreover, a higher percentage of the land is used for cash crops resulting in a stronger demand for non-agricultural products and services and therefore for cash.

For all these reasons the cropping systems have continuously changed eventually requiring the development of new IPM concepts and intensive training of farmers.

Five IPM Components

Cultural Techniques

The best method (cheap; no health hazards; environmentally friendly) to increase yields is the correct use and the improvment of cultural techniques. The importance of crop rotation in controlling pests should be pointed out here as well as positive effect of soil coverage and intercropping (Tapia Barquero & Camacho Henriques 1988).

Physical Control

In general, the physical control methods are well-adopted to the socio-economical conditions of Papua New Guinea as human labour is comparably cheap. It is a simple and - for some crops - widely-spread practice to handcollect larger-sized pests *i.e.* beetles and caterpillars. But this method is not effective in controlling small, mobile pests such as jassids or mites. Furthermore, these are often not recognised as the cause for declining yields.

Biological Methods

Biological methods being a further component of IPM have their justification in Papua New Guinea especially for introduced pests. However, their implementions require large-scale input and cooperation on the national and international level. For this reason, the introduction of biological control measures is considered to be the task of government institutions.

Chemical Control

While the disadvantages of intensive chemical control have been discussed above, it should be mentioned that the limited offer of insecticides in Papua New Guinea will force the farmers to use less specific ones. This pesticides have mostly similar chemical properties. Eventually, this will lead to a faster development of pest-resistance.

Bio-technical Control

One of the cheapest and at the same time most effective measures to control pests is the treatment of seeds. Together with basic field hygiene measures the spreading and early infection of pests and diseases are prevented. Therefore, the IPM concept for smallholders has to start with cheap, low toxic and readily available desinfection methods. This can be achieved with synthetical pesticides since the extent of environmental contamination is restricted. Furthermore, home-made plant-derived pesticides can be expected to be successfull in this respect.

As far as pheromones are concerned, bio-technical control needs a lot of financial and logistical input in both research and application, but this may change in future. Therefore, the use of bio-technical control measures in Papua New Guinea is restricted to cash crop production, *i.e.* sugar cane, oilpalm (Prior 1993; Whittle *et al.* 1992)

Advantages of Plant-Derived Pesticides

Due to their lethal as well as their repellent properties, some plant-derived pesticles (PDP) have to be regarded as "bio-technical control" rather than as "chemical control". Within the IPM concepts, they can play an important role revealing several advantages compared to chemicals.

- low persistence
- little contamination of the environment with respect to:
    - water

- non-target organisms
- accumulation in the nutrition chain
- readily available
- positive socio-cultural aspects

SOCIO-ECONOMIC BACKGROUND

In Papua New Guinea, there are two types of agricultural production: on the one hand there is the large-scale cash crop production (cocoa, coffee, coconut, sugar, oilpalm), on the other hand there are 500,000 families (Asian Development Bank 1993) of smallholder and subsistence farmers. While the former uses more or less high external inputs, the latter requires no or only very little agro-chemical input.

The IPM concepts of cash crop production in Papua New Guinea has to be considered as "stopped by half the way". The recommended use of unselective insecticides for pest control and the simultaneous use of bio-control components is contradictory and the arising problems unsolved.

From the point of IPM, the advantages of large-scale agriculture are the more professional handling of pesticides as well as the availability of capital for purchase of more selective pesticides and IPM components. Therefore, large-scale farmers can be expected to be prospective users of more sophisticated products of PDP, *i.e.* different neem (*Azadirachta indica*) products which have been produced for commercial use for some years in India, Thailand, the USA and Germany.

Nevertheless, smallholders in Papua New Guinea should be integrated in the use of PDP. After years of extensive work on PDP (Grainge & Saleem 1988; Schmutterer & Hellpapp 1989) the neem tree has evolved as the most promising plant for pest control. Its distribution is limited to the lowlands areas up to 800 m a.s.l. and rainfall should not exceed 3000 mm annually. In Papua New Guinea the neem tree is presently grown mainly in and around the capital as a shade tree and in the Markham Valley.

Probably the introduction of a related species, *Azadirachta excelsia*, which is grown already in Irian Jaya and the Philippines (Schmutterer pers. comm.) will help to spread this useful tree to more humid areas. For higher altitudes of Papua New Guinea, the neem related tree "chinaberry" or "persian lilac" (*Melia azedarach*) may be a promising plant but its properties are not as thoroughly studied as in the case of the neem tree (Zhu 1991, McMillan *et al.* 1969). Moreover, it will take some years to spread the trees equally over the country and to harvest the fruits which contain the main portion of active ingredients. Meanwhile, annual plants might help to control pests and increase both yields and income of the farmers while reducing the impact on environment and resources at the same time.

PROMISING PLANTS

The most promising agents (Schmutterer & Ascher 1986; Stoll 1992; Grainge & Salem 1988) are due to their accessibility for smallholders the use of:

chillie (Capsicum annuum),
tobacco (Tabacum rustica),
derris (Derris elliptica),
soursop (Annona spp.),
castor bean (Ricinus communis),
citronella grass (Cymbopogon naptus),
sunhemp (Crotalaria spp.);
pawpaw (Carica papaya),

tumeric (Curcuma longa) and
members of the family Aristolochiaceae

Along with some others, these plants are promising or well-known for their insecticidal properties and should be investigated with regard to simple techniques for the preparation of home-made pesticides and their efficacy to pests. Different kinds of extract methods can be compared.

It should be kept in mind that, just as synthetical pesticides, some PDP may reveal health hazards to mammals. In this case, the recommendations for a safe handling of these pesticides should be regarded.

CONCLUSIONS

The present situation for the introduction of IPM methods to small-holders in Papua New Guinea seems to be promising, since the farmers are not used to "simple solution". As chemical pesticides are very expensive and the logistical infrastructure is poor in Papua New Guinea, the distribution and use of plants known for their insecticidal properties may be an adequate solution for pest control for small-scale farming as well as for professional large-scale cash crop production.

REFERENCES

Asian Development Bank (1993). Agriculture Research and Extension project (Phase II) Papua New Guinea. Stage I report.

Brader, L. (1989). Aspect of Intergrated Pest Management in Developing Countries. *Proceedings of Intergrated Pest Management in tropical and subtropical cropping systems Vol. I.* Feb. 8-15 1989, Bad Durkheim, DLG, Germany.

Cook, K.E., Kuriata L, Ling, G, Moxon, J.E. & Nevenimo, T (1989). A Rapid Rural Appraisal of the Cocoa and Coconut based farming systems in the North East Lowlands of the Gazelle Peninsula of East New Britain Province, 2 to 14 October 1988.

FAO (1974). Proceedings of the FAO Conference Rome 1972. *Ecology in Relation to Plant Pest Control.* Publ. AGD: 1973/M/14, 326 pp.

Dinham, B. (ed. 1993). *The Pesticides hazard: a global health and environmental audit.* London : Zed Books, 228 pp.

Förster, P. (1991). Einflusse von Pflanzenschutzmitteln auf Larven und Adulte von *Platynus dorsalis* (Pont.) [Coleoptera, Carab.] und auf Adulte von *Tachyporus hypnorum* (L.) [Coleoptera: Staphylinidae] in Labor-, Halbfreiland- und Freilandversuchen. Diss. Uni Gießen.

Grainge, M. & Salem, A (1988). *Handbook of Plants with Pest Control Properties.* A. Wiley-Intersc. Publ.

Hassan, S.A. ***et al.*** **(1987).** Results of the third joint pesticides testing programme by the IOBS/WPRS Group" Pesticides and Beneficial Organisms". *J. appl. Entomol.* **103,** 92-107.

McMillan, W,W., Bowman, M.C., Burton, R.L., Starks, K.J. & Wiseman, B.R. (1969). Extract of chinaberry leaf as a feeding deterrent and growth retardent for larvae of the corn earworm and fall armyworm. *J. Econ. Ent. 62,* 708-710.

Prior, R.N.B. (1993). Limitation of IPM in cocoa subsistence farming systems and plantations in Papua New Guinea.

Schmutterer, H. (1987). *Insect growth-disrupting and fecundity reducing ingredients from the neem and chinaberry trees.* CRC Handbook of Natural Pesticides, Vol. III Insect Growth Regulators, Part B, 119-170.

Schmutterer, H. & Ascher, K.R.S. (1986). *Natural pesticides from the neem tree (*Azadirachta indica *A. Juss.) and other tropical plants.* Proceed. Third International Neem Conference, Nairobi, GTZ Schriftenreihe No. 206, Eschborn, 691 pp.

Schmutterer, H. & Hellpapp, C. (1989). Effects of neem on pests of vegetables and fruit trees. *1988-Focus on phytochemical pesticides, Vol. 1. The neem tree.* (Ed. by M. Jacobsen), pp. 69-86, Boca Raton, CRC Press.

Smith, E.H. & Pimental, D. (eds. 1978). *Pest Control Strategies.* New York, Acad. Press, 334 pp.

Stoll, G. (1989). *Natural Crop Protection in the Tropics and Subtropics.* 188 pp.

Tapia Barquero, H. & Camacho Henriques, A. (1988). *Manejo integrado de la produccion de frijol basado en labrarnaza cero.* Verlagsgesellschaft Roßdorf, GTZ Eschborn, 181 pp.

Whittle, C.P., Vickers, R.A., Kuniata, L.S., Bellas, T.E. & Rumbo, E.R. (1992). Identification of a pheromone blend attractive to Sesamia grisescens Walker [Lepidoptera: Noctuidae].

Session V: Aquatic Biology

Proceedings of the Biological Society of New Guinea
R. Höft (ed.)

BIOLOGICAL MONITORING OF AQUATIC MACROINVERTEBRATE ASSEMBLAGES IN THE FLY RIVER SYSTEM

Andrew W. Storey & Andy Y. Maie

Ok Tedi Mining Limited, P.O. Box 1, Tabubil, Western Province, Papua New Guinea

ABSTRACT

The effects of the Ok Tedi mine on the hydrology, biology and geochemistry of the Fly River system have been monitored by Ok Tedi Mining Limited since 1983. Historically, the biological monitoring program concentrated on changes in the fish fauna of the system as this constitutes a major food resource for local villagers. Macroinvertebrate assemblages were initially monitored from the early to the mid-1980's, however, further studies were deferred, with major emphasis placed on understanding and modelling the fisheries of the system. More recently the Company has implemented several new programs to monitor aquatic macroinvertebrate assemblages. This is in recognition of their likely importance in food webs and of their potential value as a tool for biological monitoring.

Three main programs are in progress; monitoring populations of Macrobrachium prawns in the Fly River, monitoring macroinvertebrate assemblages in off-river water bodies (oxbow lakes, blocked valley lakes and grassed floodplain sites) and examining rates of recovery of macroinvertebrate communities in upland creeks subjected to experimental disturbance.

The rationale for these studies, the methods utilised and preliminary results are presented and discussed.

INTRODUCTION

Before the commencement of environmental studies associated with the Ok Tedi mine there were no published data on the macroinvertebrate fauna of the Fly River system. The first recorded survey of freshwater macroinvertebrates in the system was by the 1974 Cambridge Expedition to the Western District of Papua New Guinea (Boyden et al., 1975). Approximately 16 taxa were recorded from the Ok Tedi at Tabubil, however, most specimens were identified only to family level (Table 1) and undoubtedly the total number of species would have been higher. The next major study was the comprehensive `Ok Tedi Environmental Study', which commenced in 1981 (Maunsell & Partners, 1982). Biological aspects of this work focused primarily on surveying the distribution of fish species and on recording tissue metal concentrations in fish throughout the Fly River system. There was minimal sampling of aquatic invertebrates in this study, with the exception of opportunistic plankton tows in Bossett Lagoon, Middle Fly, which recorded Volvacaceae, Copepoda and Protozoa, and the determination of tissue metal concentrations in samples of mud clams, Geloina coaxans, mud crabs, Scylla serrata and a prawn, Penaeus sp. from the Fly delta (Maunsell & Partners, 1982).

Figure 1. Locations of the freshwater sampling sites in the Fly River system

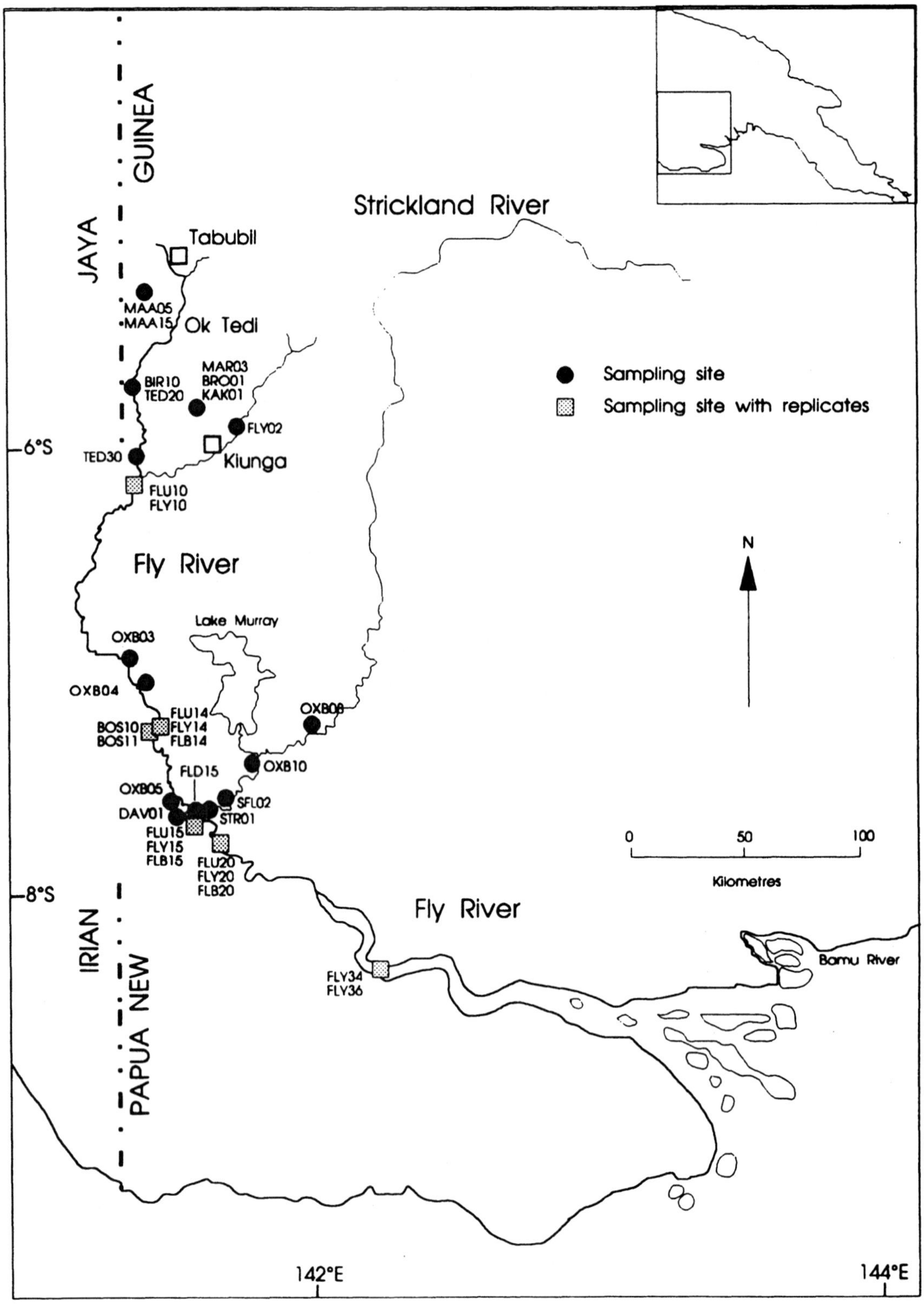

The first detailed studies of aquatic macroinvertebrate assemblages in the Fly River system were two studies, in the early to mid-1980's, by Ok Tedi Mining Limited. In early 1983 a preliminary reconnaissance survey of macroinvertebrate assemblages at 15 sites in the upper Ok Tedi catchment recorded 42 taxa (Table 1), however, due to taxonomic constraints most groups were identified only to family level (OTML, 1983). Subsequent monitoring, from mid-1983 to late 1985, noted changes in the fauna of the Ok Tedi at Tabubil, with significant reductions in density, diversity and biomass of macroinvertebrates (OTML, 1986). From early 1986 to early 1987 a comprehensive survey of the benthic fauna at nine localities in the lower Ok Tedi and Fly River recorded a total of 60 taxa (Table 1) (OTML, 1987). The fauna was dominated numerically by dipteran larvae (50 % of total numbers), however nymphs of the burrowing mayfly genus, Plethogenesia and Macrobrachium prawns dominated total biomass (OTML, 1987; 1993a).

Apart from these surveys, the primary emphasis of monitoring programs throughout the 1980's was directed at fish populations. This was essentially because fish constitute a major food resource for local villagers and therefore the Company and the State are concerned with maintaining fish stocks. Aquatic macroinvertebrate studies were deferred, with major emphasis placed on understanding and modelling the fisheries of the Fly River system.

Work on the fish of the Fly River system has resulted in a comprehensive database on approximately 95 species of fish, covering a 10 year period. Currently, 40 sites, distributed throughout the system, are sampled either on a monthly, quarterly or six monthly basis to monitor fish assemblages and tissue metal levels (Figure 1). As part of this monitoring program, fluctuations in fish biomass at a range of sites are routinely reported to the State. Trends in composition of fish assemblages with time have been investigated (Smith, in press) and the relationship between fish biomass and particulate copper concentration in the river system has been modelled (Smith & Hortle, 1991).

Ok Tedi Mining Limited is now developing programs to assess the potential effects of mining on aquatic macroinvertebrates in the Fly River system, and to determine if fish assemblages maybe influenced through processes such as "bottom-up" trophic interactions (Mills & Forney, 1988; Bartell et al., 1988). Dietary studies on the fish fauna of the Fly River system have established that macroinvertebrates form an important component in the diet of many fish species (i.e. planktivores consuming zooplankton and insectivores feeding on aquatic macroinvertebrates and adult terrestrial insects, many of which originate from the aquatic environment)(Roberts, 1978; OTML, 1986; Kare, 1991). Therefore the potential exists for a change in the macroinvertebrate fauna to affect the fish fauna. The programs will also determine the overall value of aquatic macroinvertebrates as a monitoring tool in the Fly River system. In many parts of the world (i.e. Europe, North America, Australia) aquatic macroinvertebrates have been utilised to great effect over many years to monitor changes in water quality (Hellawell, 1978; Hawkes, 1979).

Three programs are currently in progress to examine different aspects of the aquatic macroinvertebrate fauna of the Fly River System; a.) monitoring populations of Macrobrachium prawns in the Fly River, b.) monitoring macroinvertebrate assemblages in off-river water bodies (oxbow lakes, blocked valley lakes and the Fly River floodplain) and c.) investigating rates of recovery of benthic macroinvertebrate communities in upland creeks subjected to experimental disturbance.

THE STUDIES

Macrobrachium *Prawn Populations in the Fly River*

Freshwater prawns of the genus Macrobrachium together with mayfly larvae of the genus Plethogenesia constitute about 80 percent of the biomass of the macroinvertebrate fauna found in the main channel of the Fly River (OTML, 1987). Dietary studies have indicated that

Macrobrachium are an important food item for many fish species within the Fly River system (Roberts, 1978; OTML, 1986; Kare, 1991) and large species of Macrobrachium are also utilised as an important food resource by the local people. Therefore, the need to establish a specific sampling program to monitor prawns has been recognised. Macrobrachium have been sampled by seine net since the inception of biological monitoring by Ok Tedi Mining Limited. These data revealed changes in the prawn populations in the Ok Tedi following commissioning of the mining operation, with reductions in numbers during the gold operating phase but with a subsequent recovery in numbers during the copper operating phase (OTML, 1991). However, because sites suitable for seine netting (i.e. river reaches with shallow/exposed sand banks) are restricted to the upper catchment, this database is of limited value for detecting potential effects of the mine throughout the remainder of the river system.

During 1991 and 1992 a sampling methodology utilising baited traps was developed to monitor Macrobrachium prawn populations throughout the Fly River system. It has been reported previously that trap design (Yamane & Flores 1989), type of bait (Harpaz *et al.* 1987), sampling site and period (Kneipp 1979), and competitive interactions between individuals or species of Macrobrachium (Peebles 1980; Harpaz *et al.* 1987) can all influence the composition of prawn catches.

Preliminary sampling in the Fly River system, with funnel traps, identified a Macrobrachium fauna consisting of six species. Analysis of these data indicated that smaller traps caught a significantly greater proportion of smaller individuals and smaller species. These differences were assumed to be related to the funnel size of the traps and to small species avoiding traps containing large specimens.

In subsequent trials two trap types were evaluated; a hinged door box trap, similar to that of Hill (1975) and a funnel trap (Figure 2).

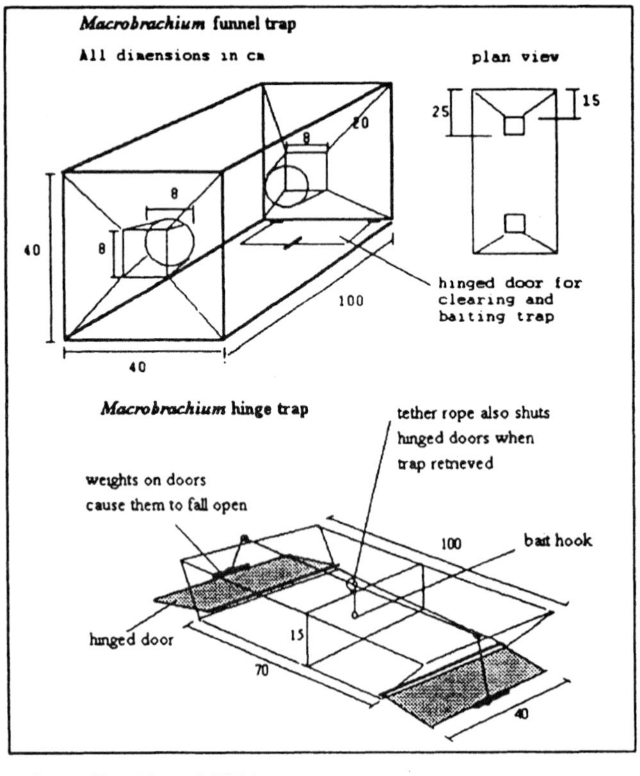

Figure 2. Schematic diagrams of the two types of prawn traps used in the study (after Tait & Bakowa, in prep.).

Five traps of each type were set alternately along the river bank at depths of 1.5 - 2.0 m. Traps were baited with freshwater herrings (Nematalosa spp.) and set for 24 hour periods. A range of trials were performed in which the frequency of checking the traps was varied from a minimum of once every 24 hours for the funnel traps to a maximum of every 45 minutes for the hinged door box traps. Trap avoidance was also tested by setting funnel traps with and without captive individuals of M. rosenbergii, a species that is known to exhibit territorial behaviour (Peebles, 1980).

The trials indicated that funnel traps, set for 24 hours and checked every 3 hours gave the optimum catches of Macrobrachium. This design increased retention of captive prawns, reduced competitive exclusion of small species and minimised logistical constraints in the field associated with frequent checking; the hinged door box traps would be effective only when checked very frequently. A detailed description of the methodology and results of this project are to be published elsewhere (Tait & Bakowa, in prep.).

A Macrobrachium monitoring program using funnel traps, modified with a 10 cm cylindrical extension to the funnel to help retain captive individuals, and based on the above methodology will be implemented at sites that are currently sampled for fish on the lower Ok Tedi, Fly and Strickland Rivers. This program will determine if the mine operation effects Macrobrachium populations at these sites.

Macroinvertebrate Assemblages in Off-River Water Bodies

The Fly River system has a catchment area of 76 000 km2 with very high rainfall, ranging from 10 000mm per annum in the upper catchment to around 3 000mm per annum near the coast. This results in a mean discharge of 6000 m3 sec-1 (OTML, 1988). Although the upper catchment extends to altitudes greater than 4000m, the majority of the catchment area is very flat. For instance, the river port of Kiunga, which is 800 river kilometres from the coast is only 20m above sea level (Smith & Bakowa, in press). The combination of high rainfall and large areas of flat land has resulted in an extensive system of wetlands. Four major habitat types are recognised in the floodplain area of the middle Fly River: blocked valley lakes, with a total area of 245km2; oxbow lakes (cut-off meander loops)(122 km2); and grassed and forested floodplain (2473 km2 combined) (Smith & Bakowa, in press).

There is little seasonality in the rainfall of the upper catchment, however the floodplain in the middle Fly is subjected to the NW monsoons (December to April) and the SE trades (May to November). As a result the water on the floodplain may be derived either from the Fly River, with associated natural and mine-derived sediments, or from local rainfall. In recent years, sites representative of the different floodplain habitats have been regularly sampled to monitor fish populations (Smith & Bakowa, in press), however, little is known of the macroinvertebrate populations in these habitats. All previous sampling of macroinvertebrate populations in the middle Fly area had been from the main river channel (OTML, 1987). Therefore, in 1993 a study was initiated to design a monitoring program that would detect any possible effects of mine-derived sediments on the macroinvertebrate fauna of floodplain habitats.

The aims of this study are to evaluate sampling methodologies for macroinvertebrates in off-river water bodies, to describe the macroinvertebrate fauna in terms of species composition, relative abundance and biomass, to determine the major habitats utilised and to develop a strategy to monitor the macroinvertebrate fauna of off-river water bodies in the Fly River system.

To achieve the stated aims, ten off-river water bodies located along the middle Fly and Strickland Rivers were selected for sampling. Replicate samples were taken from two sites within each off-river water body, providing replication between and within each waterbody. The water edge, surface and benthos were sampled using a variety of techniques: dip-net sweeps of the marginal submerged macrophytes; plankton tows in open water; Ekman-Birge grabs of the benthos in deep water; core samples of the benthos in shallow areas; and artificial substrate samples at all locations. The artificial substrates consisted of three open mesh nylon citrus bags enclosed in a coarse mesh envelope with a small rock for ballast. They were tested as a standard sampling unit which is easy to replicate and quick to process, both in the field and the laboratory. The artificial substrates were trialed over exposure periods of four, six, eight and 12 weeks to assess rates of colonization by macroinvertebrates.

This project is nearing completion of the first stage. Issues encountered so far have been the inundation or dessication of artificial substrates due to fluctuating water levels and the theft of substrates for their ropes and floats, which are an attractive resource to local villagers. At this preliminary stage 109 taxa have been identified with Coleoptera and Hemiptera as the most diverse groups. The dip-net sweep samples show a good faunal diversity but low abundances, the benthic cores and grabs reveal a patchy fauna, with low diversity and abundances and the plankton tows contained few animals. Colonization rates of the artificial substrates have yet to be examined. Once the data have been analysed an ongoing monitoring program will be initiated using the most appropriate sampling techniques for these habitats.

Benthic Macroinvertebrate Communities in Upland Creeks

During the early 1980's significant changes in the macroinvertebrate fauna of some headwater creeks in the upper Ok Tedi catchment were detected (OTML, 1986). It was postulated that sediments derived from mining activities were responsible for the observed reductions in density, diversity and biomass of macroinvertebrates in the Ok Tedi at Tabubil (OTML, 1986). Changes in the fish fauna in the Ok Tedi also have been detected, with reductions in density, diversity and biomass of fish (Smith & Morris, 1992; Smith, in press).

Mining activities, with the associated discharge of ore residues and overburden to the Ok Tedi are planned to continue into the next decade. Current modelling predicts that once mining operations cease there will be a rapid reduction in suspended sediment loads with a more gradual reduction (15-20 years) of river bed aggradation to pre-mine bed levels (OTML, 1993b). Recent sampling has demonstrated that although the number of fish species in the Ok Tedi have declined, all absent species are still found in tributary streams, or further downstream in the Fly River (Smith, in press) and it has been concluded that the fish assemblages "do not appear to have changed to compositions from which it would be difficult to return to their former states" (Smith, in press).

Similar data, however are not currently available for macroinvertebrate assemblages in the upper catchment of the Ok Tedi. Therefore, in mid-1993 a postgraduate research student from James Cook University, Queensland, Australia was funded by the Company to investigate rates of recovery of macroinvertebrate assemblages in a stream reach that was subjected to artificial disturbance. This study also would provide an inventory of the macroinvertebrate fauna in creeks unaffected by mine operations which act as refugia from which the Ok Tedi may be recolonized.

Rates of recolonization of benthic macroinvertebrates were examined by removing cobbles from the stream bed, scrubbing them clean, marking them and returning them to the same reach. This action was designed to simulate the degree of physical disturbance that occurs to a stream bed during a severe spate, with the expectation that a similar level of disturbance may result from certain mining activities. Cobbles were then removed over a period of time and the recolonization of macroinvertebrates recorded. The initial set of cobbles were removed four hours after being returned to the stream and then on every second day for an eight day period. On each occasion 'undisturbed' cobbles were also removed as a control.

Another aspect of the study looked at macroinvertebrate drift (the downstream movement of aquatic organisms in the water column) as this is recognised as a major mechanism for the recolonization of disturbed regions by macroinvertebrates (Waters, 1972). Drift samples were taken using replicate drift nets set for 24 hour periods. This would allow for the known diurnal changes in drift activity (Waters, 1969). After this period the stream bed in a 10m reach was disturbed by raking and rolling the cobbles and gravel and new drift samples were taken immediately after the disturbance and for the following 24 hour period. Cobbles were also removed from the disturbed reach and the macroinvertebrate fauna examined to gauge the magnitude of the experimental disturbance.

The field component of this project has been completed, however the identification and enumeration of the macroinvertebrate fauna has still to be completed before data analysis may commence. This initial project examined recolonization over a relatively short eight day period. In future fieldwork it is intended to examine recolonization over several months.

CONCLUSIONS

During the 1980's, studies by Ok Tedi Mining Limited on aquatic macroinvertebrates in the Fly River system were deferred in favour of intensive studies of the fish fauna. This work has been productive, both by increasing understanding of the fish fauna of the Fly River and by providing a comprehensive database which is used to manage the system. The Company is now increasing emphasis on the macroinvertebrate fauna of the system. The three research projects currently in progress will help establish effective monitoring programs to a.) detect possible effects of the Ok Tedi mine on the macroinvertebrate fauna of the Fly River system, b.) determine if the fish fauna is likely to be affected through trophic interactions and c.) provide an indication of recolonization rates for use in ecosystem management after the mine has ceased operations. This work will enhance the present understanding of the ecology of the Fly River system.

ACKNOWLEDGEMENTS

The authors thank Jim Tait and Kayemen Bakowa for the use of a summary of their investigation into sampling Macrobrachium prawns, and Steve Comport and Zoe Rosser for their assistance. Dr. R.E.W. Smith is gratefully acknowledged for his valuable contributions to the above studies during his tenure as Senior Biologist with Ok Tedi Mining Limited. Thanks also to all past and present members of staff in the Company's Biology Section who were involved in these projects and to Chris Brown, General Manager and Ian Wood, Manager Environment, for constructive criticism of the manuscript.

REFERENCES

Bartell, S.M., Brenkert, A.L., O'Neill, R.V. & Gardner, R.H. (1988). Temporal variation in regulation of production in a pelagic food web model. *Complex Interactions in Lake Communities* (Ed. by S.R. Carpenter), pp. 101-118, Springer Verlag, N.Y. Inc.

Boyden, C.R., Brown, B.E., Drucker, R.F. & Tuft, S.J. (1975). Report of the 1974 Cambridge Expedition to the Western District, Papua New Guinea, Ok Tedi Environmental Study.

Harpaz, S., Kahan, D., Galun, R. & Moore, I. (1987). Responses of the freshwater prawn, Macrobrachium rosenbergii, to chemical attractants. *J. Chem. Ecol.* **13,** 1957-1965.

Hawkes, H.A. (1979). Invertebrates as indicators of river water quality. *Biological Indicators of Water Quality* (Ed. by A. James & L. Evison), John Wiley & Sons, New York.

Hellawell, J.M. (1978). *Biological Surveillance of Rivers: A Biological Monitoring Handbook.* Water Resources Handbook, Stevenage, U.K.

Hill, B.J. (1975). Abundance, breeding and growth of the crab, Scylla serrata in two South African estuaries. *Mar. Biol.* **32,** 119-126.

Kare, B.D. (1991). *Diet of selected fish species in the Fly River system, Papua New Guinea.* Post-graduate Diploma thesis, James Cook University, Queensland. pp. 41.

Kneipp, I.J. (1979). *Ecology of* Macrobrachium *species (Decapoda: Palaemonidae) in a coastal stream in North Queensland.* PhD thesis, James Cook University, Queensland.

Maunsell & Partners (1982). Ok Tedi Environmental Study: Supporting Studies, Volume 6: Ok Tedi-Fly River Aquatic Survey-Heavy Metal Review. Report to Ok Tedi Mining Limited, pp. 344.

Mills, E.L. & Forney, J.L. (1988). Trophic dynamics and development of freshwater pelagic food webs. *Complex Interactions in Lake Communities* (Ed. by S.R. Carpenter), pp.11-30, Springer-Verlag N.Y. Inc.

OTML (1983). Reconnaissance survey of macrozoobenthos in the upper Ok Tedi catchment. OTML, Biology Report No.1, pp. 7.

OTML (1986). A review of biological sampling of the Ok Tedi and Fly River systems, April 1983 to June 1986. OTML report ENV 86-9.

OTML (1987). Studies of the benthic fauna of lowland (potamon) localities of the Ok Tedi and Fly River, with reference to mining activities. OTML, report no. ENV 87-11.

OTML (1988). Sixth Supplemental Agreement Environmental Study 1986-1988. Final draft report. Ok Tedi Mining Limited.

OTML (1991). Biology Annual Report 1991. OTML report no. ENV 91-01.

OTML (1993a). Biology Annual Report: Data collected to 30 September 1992. OTML report no. ENV 93-01.

OTML (1993b). Hydrology Annual Report 1993. OTML report no. ENV 93-03.

Peebles, J.B. (1980). Competition and habitat partitioning by the giant freshwater prawn *Macrobrachium rosenbergii* (De Man)(Decapoda, Palaemonidae) *Crustaceana* **38,** 49-54.

Roberts, T.R. (1978). An ichthyological survey of the Fly River in Papua New Guinea, with description of new species. *Smithsonian Contributions to Zoology* **281**.

Smith, R.E.W. (in press). Species diversity and trends of composition of assemblages of fish in the Fly River system, Papua New Guinea.

Smith, R.E.W. & Hortle, K.G. (1991). Assessment and prediction of the impacts of the Ok Tedi copper mine on fish catches on the Fly River system, Papua New Guinea. *Env. Mon. Assess.* **18,** 41-68.

Smith, R.E.W. & Morris, T.F. (1992). The impacts of changing geochemistry on the fish assemblages of the Lower Ok Tedi and the Middle Fly River, Papua New Guinea. *Sci. Total Environ.* **125,** 321-344.

Smith, R.E.W. & Bakowa, K.A. (in press). Utilisation of floodplain water bodies by the fishes of the Fly River. Papua New Guinea. *Mitteilungen Societas Internationalis Limnologiae*

Tait, J. T. P. & Bakowa, K.A. (in prep.). Development of a methodology to quantitatively sample *Macrobrachium* prawns in the Fly River, Papua New Guinea.

Waters, T.F. (1969). Diel patterns of aquatic invertebrate drift in streams of northern Utah. *Proc. Utah Acad. Sci.* **46,** 109-130.

Waters, T.F. (1972). The drift of stream insects. *Ann. Rev. Ent.* **17,** 253-272.

Yamane, T & Flores, E.E.C. (1989). Evaluation of the catching efficiency of small pots for prawns. *Fish. Res.* **8,** 81-91.

Table 1. Systematic listing of taxa recorded in previous studies of the aquatic macroinvertebrate fauna of the Fly River system [1 = Boyden *et al.* (1975)[1]; 2 = OTML (1983)[2]; 3 = OTML (1987)[3]]

Taxa[4] (indentation following the systematic hierarchy)						No. of Study		
NEMERTEA					Nemertea sp.			3
NEMATODA					Nematoda sp.			3
MOLLUSCA	**GASTROPODA**	PULMONATA			Gastropoda sp. A		2	
					Gastropoda sp. B		2	
					Gastropoda sp. C		2	
				Hydrobiidae	*Melanoides tuberculatus*			3
				Planorbidae	Planorbidae sp.			3
				Thiaridae	*Thiara scabra*			3
ANNELIDA	**OLIGOCHAETA**				Oligochaeta sp.		2	3
		HIRUDINEA			Hirudinea sp.			3
ARTHROPODA	**ARACHNIDA**	ORIBATIDA	HYDRACARINA		Hydracarina sp.			3
	CRUSTACEA	DECAPODA		Palaemonidae	*Macrobrachium* sp.		2	3
				Sundathelphusidae	Sundathelphusidae sp.			3
		AMPHOPODA			Amphipoda sp.			3
	INSECTA	COLLEMBOLA			Collembola sp.			3
		LEPIDOPTERA		Pyralidae	Pyralidae sp.A		2	3
					Pyralidae sp.B			3
					Pyralidae sp.C			3
					Nymphulinae sp.	1		
		DIPTERA		Simuliidae	Simuliidae sp.	1		
				Culicidae	*Chaoborus* sp.			3
				Chironomidae	Chironomidae sp.		2	3
				Tanypodinae	Anatopynia sp.	1		
				Orthocladiinae	*Orthocladius* sp.	1	2	
					Cricotopus sp.	1		
					Diamesa sp.	1		
					Procladius sp.			3
					Chironomus sp.			3
					Pentaneurini sp.			3
				Chironominae	*Polypedilum* sp.	1		
					Rheotanytarsus sp.	1		
					?Chironomini			3
					?Glyptotendipes sp.			3
					?Cardiocladius sp.			3
					?Dicrotendipes sp.			3
					Tanytarsini sp.			3
				Tipulidae	Tipulidae sp.A		2	3
					Alonia sp.			3
				Ceratopogonidae	Ceratopogonidae sp.			3
				Muscidae	Muscidae sp.			3
				Psychodidae	Psychodidae sp.		2	3
				Athericidae	Athericidae sp.		2	
				Blephariceridae	Blephariceridae sp.		2	
				Empididae	Empididae sp.		2	3
				?Tabanidae	?Tabanidae sp.	1		
				Thaumaleidae	Thaumaleidae sp.			3
		ODONATA	Zygoptera		Zygoptera sp.A		2	
					Zygoptera sp.B		2	
					Zygoptera sp.C		2	
					Zygoptera sp.D		2	
			Anisoptera		Anisoptera sp.A		2	
				Corduliidae	Corduliidae sp.		2	3
				Gomphidae	Gomphidae sp.		2	
				Synthemidae	Synthemidae sp.		2	3
		HEMIPTERA		Corixidae	Corixidae sp.		2	3
				Naucoridae	Naucoridae sp.			3
				Veliidae	Veliidae sp.		2	

[1] Ok Tedi at Tabubil (Boyden et al., 1975).

[2] 15 sites in the upper Ok Tedi catchment (OTML, 1983).

[3] 9 localities in the lower Ok Tedi and middle Fly River (OTML, 1987).

[4] Due to the absence of voucher specimens it was not possible to determine if taxa reported at family/genus level are the same taxon between the three studies.

EPHEMEROPTERA	Baetidae	Baetidae sp.A		2	
		Baetidae sp.B		2	
		Baetidae sp.C		2	
		Baetidae sp.D		2	
		Baetidae sp			3
		Baetis sp.	1		
	Caenidae	*Tasmanocoenis* sp.		2	
		Caenis sp.	1		
		Caenidae sp.A			3
		Caenidae sp.B			3
		Caenidae sp.C			3
	Leptophlebiidae	*Atalophlebia* sp.		2	
		Atalonella sp.		2	
		Leptophlebiid sp.A	1		3
		Leptophlebiid sp.B	1		
		Leptophlebiid sp.C	1		
	Palingeniidae	*Plethogenesia pallida*			3
		P. papuana			3
	Polymitarcidae	Polymitarcidae sp.		2	
	Prosopistomatidae	*Prosopistoma* sp.			3
TRICHOPTERA	Conoesucidae	Conoesucidae sp.A		2	3
		Conoesucidae sp.B		2	
		Costora sp.		2	
	Ecnomidae	*Ecnomus* sp.			3
	Glossosomatidae	*Agapetus* sp.		2	
	Hydropsychidae	*Cheumatopsyche* sp.		2	3
		Hydropsychidae sp.			3
		Hydropsychinae sp	1		
	Leptoceridae	Leptoceridae sp.	1	2	3
	Philopotamidae	Philopotamidae sp.A			3
		Philopotamidae sp.B			3
	Philorheithridae	Philorheithridae sp.		2	
	Rhyacophilidae	Apsilochorema sp.			3
		Rhyacophilidae sp.			3
COLEOPTERA	Curculionidae	Curculionidae sp.			3
	Dytiscidae	Dytiscidae sp.			3
	Elmidae	Elmidae sp.		2	3
		Austrolimnius sp.			3
		Notriolus sp.A			3
	Gyrinidae	Gyrinidae sp.		2	
	Hydrophilidae	Hydrophilidae sp.			3
	Noteridae	Noteridae sp.		2	
	Ptilodactylidae	Ptilodactylidae sp.		2	3
NUMBER OF TAXA			16	42	60

Proceedings of the Biological Society of New Guinea
R. Höft (ed.)

THE FRESHWATER FISHES OF LAKE SENTANI, JAYAPURA, IRIAN JAYA

Samuel J. Renyaan

Eastern Indonesia University Development Project IDRC/EIUDP/UNCEN

INTRODUCTION

Lake Sentani, with an area of 10.400 ha, is by far the largest lake in Irian Jaya (Howard 1987). It is situated at an elevation of 70 m, just south of the Cycloops Mountains near Jayapura.

Lake Sentani has only a single effluent, the Djafiri River which empties into the Pacific Ocean via the Tami River close to the Eastern border of Irian Jaya. The average annual rainfall in the region is 3180 to 3810 mm (FAO 1972).

According to the Inland Fishery Department Dati I, Jayapura, the greatest depth recorded in the western and the eastern parts were 140 m but only 40 m for the central part of the lake. A periodical blooming and die-off of phytoplankton has caused fish mortality, especially in the western part (November and December) and in the eastern part (May until August).

Lake Sentani is very important to the people who live around and on the lake. Fishes are caught from the lake or reared in cages, especially the Cyprinidae. To date, however, there has been no systematic study of the fish population.

The FAO 1972 study noted a fish population of 26 species that included 12 indegenous species, seven anadromous species and seven introduced species (Table 1). Howard (1987) noted 8 introduced species. According to Howard there are thus 27 species (Table 2). Allen (1991) described one more species, the Sentani Gudgeon, *Oxyleotris heterodon*. Allen's describtion of the Lake Sentani fishes were based on a preserved museum collection (Allen 1991). Exploitation of the Lake by local fishermen and stocking of the Lake by the Inland Fishery Department Dati I-Jayapura or accidently by local peoples, are resulting in changes of the fish population, including the extinction of the sawfish (FAO 1972). At present there is a lack of information concerning the species of freshwater fish to be found in Lake Sentani.

The lake also provides local people with drinking water and its beautiful scenery is an important source of inspiration for the local arts.

The prospects of further exploitation, including hydroelectric power generation by PLTA, will have an impact of unknown magnitude on the fish population. It is thus important to obtain a database as soon as possible as an inventory and stimulus to future studies. The hydroelectric power generation plant will be built by PLTA (Proyek Listrik Tenaga Air) near Yoka village, so the lake will provide Jayapura with electricity in addition it will supply drinking water and other probabilities such as recreational facilities: swimming pool, game fishing, etc.

Cenderawasih University is located in Abepura, only a few kilometres from the Lake. The Scientific Objectives of UNCEN are to develop anthropology and agriculture programs (Pola Ilmiah Pokok). The Mathematics and Science Department of the Faculty of Education UNCEN has research plans that will cover these problems:

1. Providing baseline data of Lake Sentani fishes.
2. The blooming and die-off of phytoplankton that caused fish mortalitiy. We are trying to identify the exact period of this problem so the fishermen will know what size of fish should be reared and when to harvest or remove their fish to a temporary safe place to avoid the adverse effects.

3. To identify Lake Sentani productivity.
4. To determine the abundance of fishes with higher economic value or at least valuable to the local people, such as the Ophiocephalidae.

METHODS

Freshwater fishes of Lake Sentani were collected from: (1) Cages around the lake, (2) Fishermen who use either passive or active capture methods, (3) Stations which were determined according to diversity of habitats in the lake (Nielsen and Johnson 1983).

The freshwater fishes were classified into 5 main categories: (1) Fish reared in cages, (2) Fish endemic to the lake, (3) Fish introduced into the lake, (4) Aquarium fishes living in the lake, (5) Fish reported from the lake according to the literature, but not collected during the sampling periods.

Fish collected was identified and classified into family, genus and species (Munro 1967; Saanin 1968; Department of Agriculture 1969; Allen & Cross 1982; Axelrod 1987; Allen 1991). Sampled fish was preserved in formaldehyde and photographs were taken for further identification.

Identification was mostly based on colour descriptions and diagnostic features.

COLLECTING STATIONS

Stations were determined according to Lake Sentani's regions. There are 3 sampling regions with 5 collecting sites: (1) Eastern Sentani with Yoka port and Puai, (2) Central Sentani with Yakhim port and Sentani Market, and (3) Western Sentani with only Boroway port.

OBJECTIVES

1. To make a database of Lake Sentani fishes.
2. To make identification keys of Lake Sentani fishes.
3. To introduce Lake Sentani fish species into the study programs of students at UNCEN.

DESCRIPTION OF LAKE SENTANI'S FRESHWATER FISHES (SYSTEMATICS)

The species of freshwater fishes in Lake Sentani has increased if compared with what FAO (1972) and Howard (1987) reported. A full description of each species (systematics) and pictures of collected fishes will be provided in the final report. Here, data is presented in Tables 1, 2 and 3.

Table 1. The freshwater fishes of Lake Sentani (Modified from FAO, 1972).

Group (1) (2) (3) (4)	Family	Specific Name and Local Name
A.Indigenous	a.Tachysuridae	1.*Hemipimelodus velutinus* Kanseli, species Weber Sembilang
	b.Plotosidae	2.*Neosilurus novaeguineae* Holiya Weber
	c.Eleotridae	3.*Oxyeleotris lineolatus* Humen, Steindachner Gabus
		4.*Bunaka herwedeni* Gabus Weber
		5.*Ophiocara aporos* Blkr. Humem, Gabus
		6.*Pogoneleotris microps* Humem, Weber Gabus
	d.Gobiidae	7.*Glossogobius giurus* Humem, Ham.-Buch. Gabus
	e.Apogonidae	8.*Apogon wichmani* Kahe, Weber Gete-gete
		9.*Apogon beauforti* Kandei, Weber Gete-gete
	f.Atherinidae	10.*Chilatherina sentaniensis* Hewa, Weber Kaskado
		11.*Glossolepis incisus* Hewu, Weber Kaskado
B.Anadromes	g.Anguillidae	12.*Anguilla australis* Kahilo, Species Weber Belut
	h.Pristidae	13.*Pristis microdon* Merlemai, Latham Gergaji
	i.Carangidae	14.*Caranx stellatus* Barra Eydoux et Souleyet
		15.*Caranx ignobilis* Barra Forskal
	j.Mugilidae	16.*Mugil cephalus* Kaijoko, Linnaeus Belanak
	k.Elopidae	17.*Megalops cyprinoides* - Broussonet
	l.Lutjanidae	18.*Lutjanus sp.* -
	m.Chanidae	19.*Chanos chanos* Forskal Bandeng
C.Introduced	n.Cyprinidae	20.*Cyprinus carpio* Mas Species Linnaeus
		21.*Puntius gonionotus* Tawes Bleeker
		22.*Puntius orphoides* Mata merah Cuv. & Val.
	o.Anabantidae	23.*Helostoma temmincki* Tambakan Cuvier
		24.*Trichogaster pectoralis* Sepat Siam Regan
		25.*Osphronemus goramy* Gourami Lacepede
	p.Cichlidae	26.*Tilapia mossambica* Mujair Peters

Table 2. The freshwater fishes of Lake Sentani (Modified from Howard, 1987)

Group (1) (2) (3) (4)	Family	Specific Name Local Name
A.Indigenous	a.Tachysuridae	1.*Hemipimelodus velutinus* Kanseli, species Weber Sembilang
	b.Plotosidae	2.*Neosilurus novaeguineae* Holiya Weber
	c.Eleotridae	3.*Oxyeleotris lineolatus* Humen, Steindachner Gabus
		4.*Bunaka herwedeni* Gabus Weber
		5.*Ophiocara aporos* Blkr. Humem, Gabus
		6.*Pogoneleotris microps* Humem, Weber Gabus
	d.Gobiidae	7.*Glossogobius giurus* Humem, Ham.-Buch. Gabus
	e.Apogonidae	8.*Apogon wichmani* Kahe, Weber Gete-gete
		9.*Apogon beauforti* Kandei, Weber Gete-gete
	f.Atherinidae	10.*Chilatherina sentaniensis* Hewa, Weber Kaskado
		11.*Glossolepis incisus* Hewu, Weber Kaskado
B.Anadromes	g.Anguillidae	12.*Anguilla australis* Kahilo, Species Weber Belut
	h.Pristidae	13.*Pristis microdon* Merlemai, Latham Gergaji
	i.Carangidae	14.*Caranx stellatus* Barra Eydoux et Souleyet
		15.*Caranx ignobilis* Barra Forskal
	j.Mugilidae	16.*Mugil cephalus* Kaijoko, Linnaeus Belanak
	k.Elopidae	17.*Megalops cyprinoides* - Broussonet
	l.Lutjanidae	18.*Lutjanus sp.* -
	m.Chanidae	19.*Chanos chanos* Forskal Bandeng
C.Introduced	n.Cyprinidae	20.*Cyprinus carpio* Mas Species Linnaeus
		21.*Puntius gonionotus* Tawes Bleeker
		22.*Puntius orphoides* Mata merah Cuv. & Val.
	o.Anabantidae	23.*Helostoma temmincki* Tambakan Cuvier
		24.*Trichogaster pectoralis* Sepat Siam Regan
		25.*Osphronemus goramy* Gourami Lacepede
	p.Cichlidae	26.*Oreochromis mossambica* Mujair Peters
		27.*Oreochromis nilotica* Mujair

Table 3. The freshwater fishes of Lake Sentani (Collected and reported by Renyaan, 1993).

Group (1) (2) (3) (4)	Family	Specific Name Local Name
A.Indigenous	a.Tachysuridae	1.*Arius velutinus* Kanseli, species Sembilang
	b.Plotosidae	2.*Neosilurus novaeguineae* Holiya Weber
	c.Clariidae	3.*Clarias batrachus* Lele Weber & de Beaufort
	d.Eleotridae	4.*Oxyeleotris heterodon* Humen Steindachner Gabus
		5.*Oxyeleotris herwedinii* Gabus Weber
		6.*Ophieleotris aporos* Humem, Blkr. Gabus
	e.Channidae	7.*Channa striata* Gabus Bloch.
	f.Apogonidae	8.*Apogon wichmani* Kahe, Weber Gete-gete
		9.*Apogon beaufori* Kandei, Weber Gete-gete
	g.Atherinidae	10.*Chilatherina sentaniensis* Hewa, Weber Kaskado
		11.*Glossolepis incisus* Hewu, Weber Kaskado
B.Anadromes	h.Anguillidae	12.*Anguilla australis* Kahilo, Species Weber Belut
		13.*Anguilla bicolor* Belut McCleland
		14.*Anguilla marmorata* Belut Quoy & Garmand
		15.*Anguilla obscura* Belut Gunther
	i.Pristidae	16.*Pristis microdon* Merlemai, Latham Gergaji
	j.Carangidae	17.*Caranx stellatus* Barra Eydoux et Souleyet
		18.*Caranx ignobilis* Barra Forskal
	k.Mugilidae	19.*Mugil cephalus* Kaijoko, Linnaeus Belanak
C.Introduced	l.Cyprinidae	20.*Cyprinus carpio* Mas Species Linnaeus
		21.*Cyprinus spp.* Mas
		22.*Puntius gonionotus* Tawes Bleeker
		23.*Puntius orphoides* Mata merah Cuv. & Val.
		24.*Osteochilus hasseltii* Nilem hijau Cuv. & Val
		25.*Osteochilus sp.* Nilem merah
	m.Anabantidae	26.*Helostoma temmincki* Tambakan Cuvier
		27.*Trichogaster pectoralis* Sepat Siam Regan
		28.*Osphronemus goramy* Gourami Lacepede
	n.Cichlidae	29.*Oreochromis mossambica* Mujair Peters
		30.*Oreochromis spp.* Mujair
		31.*Oreochromis nilotica* Mujair
	o.Cyprinodontidae	32.*Xiphophorus sp.* Ekor pedang

CONCLUSIONS AND IMPLICATIONS

1. The fish population species continuously changes from 26 species (FAO, 1972) to 27 species (Howard, 1987) and now to 32 species (Renyaan, 1993).
2. Preparing a database of Lake Sentani Fishes
3. Monitoring Lake Sentani fish (species abundancy).
4. Recording the occurence of economically important fish for further study, such as determining their abundance.
5. Preparing identification keys for Lake Sentani fishes.
6. Teaching about Lake Sentani fishes as local fish resources to Biology students.

REFERENCES AND RELEVANT LITERATURE

Allen, G. R. (1991). *Field Guide to the Freshwater Fishes of New Guinea.* Christensen Research Institute. Madang. Papua New Guinea. 268 pp.

Allen, G. R. & Cross, N.J. (1982). *Rainbowfishes of Australia and Papua New Guinea.* T.F.H. Publications, Inc. Neptune City, NJ. 141 pp.

Allen, G. R. & Coates, D. (1990). An Ichthyological Survey of the Sepik River, Papua New Guinea. *Rec. Western Australian Museum.* Suppl. No. **34,** 31-116.

Allen, G. R. & Hoese, D. F. (1980). A Collection of Fishes from the Jardine River, Cape York Peninsula. *Australia. J. R. Soc. West. Aust.* **63(2),** 53-61.

Axelrod, H. R. (1987). *Dr. Axelrod's Mini Atlas of Freshwater Aquarium Fishes.* T.F.H. Publications, Inc. Neptune, NJ 992 pp.

Chambers, M.R. (1987). The Freshwater Lakes of Papua New Guinea: An Inventory and Limnological Review. *J. Trop. Ecol.* **3(1),** 1-23.

Coates, D. (1985). Fish Yield Estimates for the Sepik River Papua New Guinea, A large Floodplain System East of Wallace Line. *J. Fish Biol.* **27(4),** 431-444.

Coates, D. (1986). Sepik River Fishery Research 1981 to 1984: Description of Methods and Raw Data. Tech. Rep. Fish. Div. Port Moresby. Papua New Guinea. No. 86-5.

Coates, D. (1986). Fisheries development of the Sepik River, Papua New Guinea. The First Asian Fisheries Form. Proceedings of the First Asian Fisheries Forum. Manila. Philippines, pp. 367-370.

Coates, D. (1989). *Papua New Guinea. A review of aquaculture and freshwater fisheries in Papua New Guinea.* FAO. Rome Italy. 30 pp.

Departement of Agriculture (1969). *A Guide to the Field Identification of the Freshwater fish and Crustacea of New Guinea.* Departement of Agriculture, Stock and Fisheries. Port Moresby, New Guinea. 15 pp.

FAO (1972). *Inland Fisheries Development in West Irian.* FAO, Rome. 41 pp.

Haines, A.K. (1979). An Ecological Survey of the Fish of the Lower Purari River System, Papua New Guinea. Office of Environment and Conservation. Waigini. Papua New Guinea. P. Environ. Stud. Purari River (Wabo). Hydroe lectr. Scheme, V. 6.

Haines, A.K. (1983). Fish Fauna and Ecology. In the Purari: Tropical Environment of a High Rainfall River Basin. Petr. T.. *Monogr. Biol.* **51,** 367-384.

Howard, K.T. (1987). *Inland Fisheries Development Irian Jaya.* Government of the Republic of Indonesia. 89 pp.

Kailola, P. J. (1987). *The Fishes of Papua New Guinea. A revised and annoted Checklist. Volume I and II.* Research Bull., Department Fish and Marine Resources. Port Moresby. 418 pp.

Munro, I. S. R. (1967). *The Fishes of New Guinea.* Departement of Agriculture, Stock and Fisheries, Port Moresby, New Guinea. 651 pp.

Nielsen, L.A. & Johnson, D.L. (1983). Fisheries Technic. AFS Bethesda, MA. 468 pp.

Osborne, P.L.; Kyle, J.H. & Abramski, M.S. (1987). Effects of Seasonal Water Level Changes on the Chemical and Biological Limnology of Lake Murray, Papua New Guinea. *Aus. J. Mar. Freshwat. Res.* **38(3),**.397-408.

Roberts, T. R. (1978). An Ichthyological Survey of the Fly River in Papua New Guinea with Descriptions of New Species. *Smithsonian Contrib. Zool.* **281,** 1-72.

Saanin, H. (1968). *Taksonomi dan Kuntji Identifikasi Ikan. Jilid I dan II.* Penerbit Binatjipta, Bandung. 265 pp.

SESSION VI: LAND VERTEBRATES

Proceedings of the Biological Society of New Guinea
R. Höft (ed.)

THE DIET OF FIVE SPECIES OF NEW GUINEA RODENTS

Karen Jackson & Patricia A. Woolley

Department of Zoology, La Trobe University Bundoora, Victoria 3083, Australia

ABSTRACT

The gut structure of *Lorentzimys nouhuysi*, *Mayermys ellermani*, *Neohydromys fuscus* and *Pseudohydromys mutinus* has been examined and compared with extra-limital species known to be strictly herbivorous or strictl carnivorous. Gut contents have been analysed. It is concluded that these five species are adapted to a mixed diet with a trend towards more herbivory in *Lorentzimys* and towards carnivory in the others with *Pseudohydromys murinus* the most carnivorous.

Proceedings of the Biological Society of New Guinea
R. Höft (ed.)

FRUIT BATS ON MANSINAM ISLAND, IRIAN JAYA

Matheus St. Kilmaskossu

INTRODUCTION

Mansinam is a small island located off Manokwari. It forms a natural shelter to the harbour of Manokwari, covering an area of about 1,855 hectares and can be reached easily within 15 minutes from Manokwari using a motorboat. On this island Ottow and Geisler, two German Evangelists landed on February 5, 1885 and for the first time spread the Christian religion in Irian Jaya.

On this well-known island is also found a large fruit bat population. They migrated tothe island four or five years ago since their former native living places on the mainland had been disturbed due to forest exploitation and hunting activities. The island is still covered with primary rainforest and secondary forest.

As part of the forest ecosystem, fruit bats play an important role in forest regeneration both as pollinators and seeds distributors. Based on a survey and several observation, this paper tries to expose and discuss the existing bat populations on Mansinam that migrate to the island to find a shelter for their survival. As the bats had been disturbed lately, this paper also tries to seek for ways of preservation as an effort to sustain the forest ecosystem.

METHODS

The method used is a descriptive method with a survey technique. To know the bat populations, their habitats and movements, a survey had been conducted in the first week of July 1993 and several observation have been done since May 1993.

Ten forestry students were enrolled in this survey. A transect line had been laid for a vegetation analysis and the existing vegetation identified. Some bats had been caught or bought from the villagers or hunters to be identified and measured.

Supporting data had been gathered by interviewing several villagers while secondary data were gathered from the village office.

RESULTS AND DISCUSSION

The Description of Bats on Mansinam

The fruit bats living on Mansinam belong to the species of *Pteropus neohibernicus papuanus* Peters, and consist or can be divided further into three subspecies based on physical appearance. These are namely: the black type, the brown breast type and the white breast type.

Their body weight ranges from 500 grams to 2500 grams. The maximum length of the wings reaches 160 cm and the width 30 cm. The body length ranges from 19 to 28 cm and the breast width 15 cm. Both the brown breast and the white breast bats have a white circle around their neck or are ring-necked, but they can be easily recognized by the striking different colour of their breast. These three bat subspecies live mixed together.

The Habitat and Life activity

The bats on Mansinam live in colonies on big tall trees. The major tree species the bats are hanging in and resting during the day time are *Pometia* sp., *Intsia* sp. and *Ficus* sp. These trees are preferred by the bats due to their big and strong branches to cling on and besides that they are the tallest trees on this island (20 to 40 metres) high.

Before dark or sunset the bats will fly out to look for food. Most of them fly to Maruni and Kacarro on the mainland as far as 3 to 4 km and come back to their resting trees before sunrise.

The fruits the bats feed on or use as their daily menu consist of Ketapang fruits (*Terminalia catappa*), bananas (*Musa* spp.), Pawpaw (*Carica papaya*), Mangga (*Anacardium* spp.), Jambu (*Eugenia* spp.), Bread fruit (*Artocarpus* spp.) and many other wild fruit trees.

In certain months, during the fruiting season of Ketapang, the bats are not flying across the bay to the main land but prefer to stay and consume the Ketapang fruits which are plentiful on Mansinam.

Hunting activities and their impact

Since their migration from the mainland to this island about five years ago, the bat populations seems to increase in number. The birth rate is high thanks to their habitat.

The local community or villagers using bows and arrows shoot no more than 3 or 4 individuals of bats for their own consumption. Lately, many hunters from the city of Manokwari came to Mansinam to shoot the bats. Using air rifles of callibre 4.5 a hunter can collect more than 30 bats. Since the bats have good strong and sharp nails they can hang tightly at a twig even after being killed. To get down the killed bats, the hunters cut down the *Pometia*, *Intsia* and the surrounding trees.

It is not a surprise to notice that the bats are now migrating to the mainland to a different place since they are threatened and their habitat is being destroyed. Besides being disturbed by the hunters, the lumbermen come and cut the host trees to harvest the timber.

As we know, *Pometia* and *Intsia* trees are endemic in Irian Jaya and are the dominant species that occupy the upper strata. They have a high quality as timber for construction. These two tree species together with other commercial timber have been exploited during the last two decades. Even though Mansinam and its forest does not belong to any concession holders, people or small entrepreneurs are going to buy or cut those trees.

By felling the *Pometia* and *Intsia* trees both the habitat of bats and the gene resources may be destroyed.

The bat problem and the hunting, as has been exposed above, give us the picture of how sensitive and vulnerable the tropical rain forest is to all kinds of disturbances. To get a killed bat down the hunters fell down the tree. Similarly the lumber men cut the tree without paying attention to the existing wildlife that is living in or at the tree and that their life depends very much on that tree. The case of forest ecosystem disturbance on Mansinam is just a very small part of forest destruction which is occurring or happening in other parts of Indonesia or even in other parts of the world.

The Future Plan

Due to the hunting activities accompanied with timber harvesting the bats population seems to decrease again. Based on our observations, it seems that the fluctuation depends very much on hunting activities. For example in May 1993 we encountered 7 colonies. At the moment we assume that a group or several groups occupying a single tree is one colony. At the end (fourth week) of July only one colony was met. But two weeks ago (mid August 1993) we found 5 colonies again. The bats become more shy.

Since little is known about bats a lot needs to be studied about bats as part or element of the ecosystem. First of all, we should manage to keep them on Mansinam. In other words, we should make Mansinam a "Home Sweet Home" for the bats to stay.

When we can manage to keep them on Mansinam then it will be easy to conduct studies or research concerning bats since they are now easy to reach or to get. In order to achieve this goal some action needs to be taken among other restriction of hunting activities, prohibition of cutting big *Pometia* and *Intsia* trees, extension and information should be given to the people etc.

CONCLUSIONS

The fruit bats living on Mansinam island belong to the species of *Pteropus neohibernicus papuanus* Peters. Their population fluctuates following the hunting activities.

Both hunting and timber cutting affect the ecosystem direct or indirectly by altering the habitat of wildlife depending on the trees and so decreasing the gene resources .

To maintain the bats colonies and keep them on Mansinam for further studies by improving the habitat, regulations should be launched and enforced.

Proceedings of the Biological Society of New Guinea
R. Höft (ed.)

SURVEY OF THE AVIFAUNA OF KUPER RANGE, MOROBE PROVINCE, PAPUA NEW GUINEA

Martin Hutten, Philippe Cuénoud [†] & Robert Höft [‡]

Duistervoordseweg 62, 7391 CG Twello, The Netherlands
[†] *11B, Ch. des Semailles, 1212 Gd-Lancy, Switzerland*
[‡] *Wau Ecology Institute, P.O. Box 77, Wau, Morobe Province, Papua New Guinea*

ABSTRACT

Introduction

The avifauna in the vicinity of the Wau Ecology Institute field station in the Kuper Range was surveyed during a total of 36 observation days. Between April and June 1993 sightings of birds were recorded and bird species were tentatively categorised according to their abundance and frequency.

Study Area

The area investigated during the study has been described in some detail by Kiapranis & Lawong (this volume). The study was carried out in an altitudinal range between 1850 and 2050 m a.s.l. The vegetation consisted of mostly undisturbed mid-montane moss forest dominated by members of the Fagaceae family. The bulk of the observations was made along a road cut resulting in an open strip of up to 50 m width. The road sides were dominated by dense tangles of secondary vegetation: pioneer shrubs, climbing bamboos and leftover trees which were frequently covered with vines offering a variety of habitats and food sources.

Methods

In the resulting list (Appendix 1), one of four categories of a subjective frequency/abundance scale were assigned to every bird species:

- U = status uncertain(few sightings)
- I = irregular frequency (few scattered sightings or presence periodical)
- R = regular(recorded on more than 10 percent of the survey days)
- C = common (recorded on more than 20 percent of the survey days)

Most of the observations were made by the first author but some sightings by other persons have been included if the identification could be confirmed.

Results

A total of 99 species was recorded. Among the interesting records are a sighting of the Yellowish Streaked Honeyeater *Ptiloprora meekiana*, a rare first bird which had not been recorded from the area and the presence of the Black Sicklebill *Epimachus fastuosus*. A striking feature was the significant increase in the abundance of several parrot species when the ericaceous vine *Dimorphanthera kempteriana* Schltr. was in flower.The species which were observed feeding on the flowers included Goldie's Lorikeet (*Trichoglossus goldiei*), the Plum-faced Lorikeet (*Oreopsittacus arfaki*), the Dusky Lory (*Pseudeos fuscata*), the Papuan Lorikeet (*Charmosyna papou*), the Little Red Lorikeet (*Charmosyna pulchella*) and the Pygmy Lorikeet (*Charmosyna wilhelminae*). The highly nomadic behaviour of some species (Goldie's lorikeet, Blue-collared Parrot, *Geoffroyus simplex*, Papuan Mountain Pigeon, *Gymnophaps albertisii*) was conspicuous, these species being abundant during some periods whilst virtually absent during others. Moreover, it was surprising to observe some flocks of Dusky Lories (*Pseudeos fuscata*), a species which is frequent in disturbed submontane areas, and whose altitudinal range rarely reaches 1800 m (Beehler *et al.* 1986). This sighting can be interpreted as nomadic behaviour as well, the observed birds having been present for two days only (during the bloom of *Dimorphanthera*).

Discussion

Despite the limited time of surveying, which would not allow any statements on the seasonality of birds, it is hoped that the list presented here may provide a good basis for any birdwatcher visiting the field station of Wau Ecology Institute.

REFERENCES

Beehler, B.M., Pratt, T.K. & Zimmermann, D.A. (1986). *Birds of New Guinea.* Wau Ecology Handbook No. 9, Princeton University Press, 294 pp.

Kiapranis, R.& Lawong, B. (this volume). Forest structure of a montane rain forest in Wau, Morobe Province, Papua New Guinea

APPENDIX 1

List of bird sightings in the Kuper Range, Morobe Province with categories of frequency/abundance (U = status uncertain, I = irregular frequency, R = regular, C = common); the first number between the brackets indicates how many individual birds were seen. The second gives the total number of days the particular bird was observed. More or less extensive bird recordings have been made on in total 36 days.

CASUARIIDAE:	CASSOWARIES	
Casuarius bennetti	Dwarf Cassowary	I(1-1)
ACCIPITRIDAE:	HAWKS, KITES AND EAGLES	
Henicopernis longicauda	Long-tailed Buzzard	R(5-4)
Accipiter melanochlamys	Black-mantled Goshawk	U(1-1)
Hieraaetus morphnoidus	Little Eagle (T.R.)	U(1-1)
Harpyopsis novaeguineae	New Guinea Harpy Eagle	U(2-2)
MEGAPODIIDAE:	MEGAPODES OR MOUND-BUILDERS	
Megapodius freycinet	Common Scrubfowl	U(4-3)
Talegalla jobiensis	Brown-collared Brush-turkey	U(2-2)
Aepypodius arfakianus	Wattled Brush-turkey	U(1-1)
RALLIDAE:	RAILS, MOORHENS AND COOTS	
Rallina forbesi	Forbes' Forest-Rail	U(3-1)
COLUMBIDAE:	PIGEONS AND DOVES	
Reinwardtoena reinwardtii	Great Cuckoo-Dove	U(1-1)
Macropygia nigrirostris	Black-billed Cuckoo-Dove	I(5-2)
Gallicolumba jobiensis	White-bibbed Ground-Dove	U(5-1)
Gallicolumba beccarii	Bronze Ground-Dove	U(1-1)
Ptilinopus rivoli	White-breasted Fruit-Dove	R(5-4)
Ptilinopus ornatus	Ornate Fruit-Dove (E race)	U(3-2)
Ducula chalconota	Rufescent Imperial Pigeon	C(11-7)
Columba vitiensis	White-throated Pigeon	U(1-1)
Gymnophaps albertisii	Papuan Mountain Pigeon	C(326-12)
Otidiphaps nobilis	Pheasant Pigeon	I(3-3)
PSITTACIDAE:	PARROTS, LORIES AND COCKATOOS	
Pseudeos fuscata	Dusky Lory	U(14-2)
Trichoglossus goldiei	Goldie's Lorikeet	I(67-4)
Charmosyna pulchella	Little Red Lorikeet	I(10-4)
Charmosyna wilhelminae	Pygmy Lorikeet	U(4-2)
Charmosyna papou wahnesi	Papuan Lorikeet (bl + r phase)	C(86-14)
Neopsittacus pullicauda	Orange-billed Lorikeet	R(15-5)
Neopsittacus musschenbroekii	Yellow-billed Lorikeet	C(27-11)
Oreopsittacus arfaki	Plum-faced Lorikeet	U(10-2)
Psittacella madaraszi	Madarasz's Tiger-Parrot	C(9-7)
Ptittacella brehmii	Brehm's Tiger-Parrot	C(12-7)
Alisterus chloropterus	Papuan King-Parrot	C(17-13)
Geoffroyus simplex	Blue-collared Parrot	I(32-5)
CUCULIDAE:	CUCKOOS, KOELS AND COUCALS	
Cacomantis castaneiventris	Chestnut-breasted Cuckoo	U(3-2)
Cacomantis flabelliformis	Fan-tailed Cuckoo	U(1-1)
Chrysococcyx ruficollis	Rufous-throated Bronze-Cuckoo	U(2-2)
TYTONIDAE:	BARN OWLS	
Tyto tenebricosa	Sooty Owl	U(1-1)
HEMIPROCNIDAE:	TREE-SWIFTS	
Hemiprocne mystacea	Moustached Tree-swift (T.R.)	U

APODIDAE:	SWIFTS AND SWIFTLETS	
Collocalia hirundinacea	Mountain Swiftlet	U(1-1)
Çollocalia esculenta	Glossy Swiftlet	C(10-10)
ALCEDINIDAE:	KINGFISHERS AND KOOKABURRAS	
Halcyon megarhyncha	Mountain Kingfisher	U(2-2)
Halcyon macleayii	Forest Kingfisher (P.C.)	U(1-1)
Clytoceyx rex	Shovel-billed Kingfisher (T.R)	U
MEROPIDAE:	BEE-EATERS	
Merops ornatus	Rainbow Bee-eater	U(1-1)
CAMPEPHAGIDAE:	CUCKOO-SHRIKES AND TRILLERS	
Coracina longicauda	Hooded Cuckoo-shrike	C(10-9)
Coracina caeruleogrisea	Stout-billed Cuckoo-shrike	I(9-3)
Coracina montana	Black-bellied Cuckoo-shrike	U(1-1)
ORTHONYCHIDAE:	LOGRUNNERS AND ALLIES	
Ptilorrhoa leucosticta	Spotted Jewel-babbler	U(3-2)
Ifrita kowaldi	Blue-capped Ifrita	C(14-7)
SYLVIIDAE:	OLD WORLD WARBLERS	
Phylloscopus trivirgatus	Island Leaf-Warbler	U(3-3)
MALURIDAE:	FAIRY-WRENS	
Clytomyias insignis	Orange-crowned Fairy-wren	U(7-3)
ACANTHIZIDAE:	AUSTRALIAN WARBLERS	
Crateroscelis robusta robusta	Mountain Mouse-warbler	C(12-8)
Sericornis arfakianus	Buff-faced Scrub-wren	R(7-4)
Sericornis nouhuysi	Large Scrub-wren	C(27-12)
Sericornis arfakianus	Grey-green Scrub-wren	U(3-2)
Gerygone ruficollis	Brown-breasted Gerygone	U(3-1)
Gerygone cinerea	Grey Gerygone	U(7-1)
RHIPIDURIDAE:	FANTAILS	
Rhipidura brachyrhyncha	Dimorphic Fantail (d + l ph)	C(11-8)
Rhipidura atra	Black Fantail	C(9-7)
Rhipidura albolimbata	Friendly Fantail	C(16-16)
MYIAGRIDAE:	MONARCHS	
Monarcha axillaris	Black Monarch	U(3-2)
Machaerirhynchus nigripectus	Black-breasted Boatbill	C(6-6)
EOPSALTRIIDAE:	AUSTRALIAN ROBINS	
Microeca papuana	Canary Flycatcher	C(7-7)
Poecilodryas albonatata	Black-throated Robin	C(6-5)
Poecilodryas albispecularis	Ashy Robin	U(1-1)
Peneothello cyanus	Blue-grey Robin	C(13-11)
Tregellasia leucops	White-faced Robin	U(1-1)
PACHYCEPHALIDAE:	WHISTLERS, PITOHUIS AND ALLIES	
Pachycephala rufinucha	Rufous-naped Whistler	I(5-4)
Pachycephala soror	Sclater's Whistler	U(3-3)
Pachycephala schlegelii	Regent Whistler	C(29-19)
Pachycephala modesta	Brown-backed Whistler	I(4-3)
Pitohui nigrescens	Black Pitohui	U(3-2)
Colluricincla megarhyncha	Little Shrike-thrush	U(1-1)
Eulacestoma nigropectus	Wattled Ploughbill	R(9-5)
CLIMACTERIDAE:	AUSTRALIAN TREECREEPERS	
Cormobates placens	Papuan Treecreeper	C(11-7)

NEOSITTIDAE:	SITTELLAS	
Daphoenositta chrysoptera	Varied Sitella	C(34-7)
DICAEIDAE:	FLOWERPECKERS AND BERRYPECKERS	
Oreocharis arfaki	Tit Berrypecker	R(5-4)
Melanocharis versteri	Fan-tailed Berrypecker	U(2-1)
Melanocharis longicauda	Mid-mountain Berrypecker	C(11-7)
Dicaeum pectorale	Papuan Flowerpecker	R(4-4)
ZOSTEROPIDAE:	WHITE-EYES	
Zosterops novaeguineae	New Guinea White-eye	C(195-9)
MELIPHAGIDAE:	HONEYEATERS	
Myzomela rosenbergii	Red-collared Myzomela	C(19-13)
Meliphaga orientalis	Mountain Meliphaga	U(6-1)
Pycnopygius cinereus	Marbled Honeyeater	I(4-1)
Melipotes fumigatus	Common Smoky Honeyeater	C(17-16)
Melidectes belfordi	Belford's Melidictes (hybr)	C(18-17)
Ptiloprora guisei	Rufous-backed Honeyeater	C(11-10)
Ptiloprora meekiana	Yellowish Streaked Honeyeater	U(4-3)
Lichenostomus subfrenatus	Black-throated Honeyeater	C(24-11)
ESTRILDIDAE:	MANNIKINS, PARROT-FINCHES	AND ALLIES
Erythrura trichroa	Blue-faced Parrot-Finch	R(10-5)
Erythrura papuana	Papuan Parrot-finch	I(3-3)
DICRURIDAE:	DRONGOS	
Chaetorhynchus papuensis	Mountain Drongo	U(2-1)
GRALLINIDAE:	MAGPIE-LARKS	
Grallina bruijni	Torrent-lark	U(1-1)
ARTAMIDAE:	WOOD-SWALLOWS	
Artamus maximus	Great Wood-swallow	C(38-7)
CRACTICIDAE:	BUTCHERBIRDS AND ALLIES	
Peltops montanus	Mountain Peltops	R(7-5)
PTILONORHYNCHIDAE:	BOWERBIRDS	
Amblyornis macgregoriae	Macgregor's Bowerbird	I(4-3)
PARADISAEIDAE:	BIRDS OF PARADISE	
Lophorina superba	Superb Bird of Paradise	U(2-2)
Astrapia stephaniae	Stephanie's Astrapia	C(29-17)
Epimachus meyeri	Brown Sicklebill	C(12-10)
Parotia lawesii	Lawes' Parotia	U(2-2)
Manucodia keraudrennii	Trumpet Manucode	U(1-1)

Abbreviations: (r + bl ph) = red + black phase, (d + l ph) = dark + light phase; (hybr) = hybrid form

SESSION VII: SYSTEM ECOLOGY

Proceedings of the Biological Society of New Guinea
R. Höft (ed.)

THE REGENERATION POTENTIAL OF A SUBMONTANE '*KUNAI*' GRASSLAND SITE IN WAU, MOROBE PROVINCE, PAPUA NEW GUINEA.

Robert Höft, Sigrun Lange[†] & Otto Spang[†]

Wau Ecology Institute, P.O. Box 77 Wau, Morobe Province, Papua New Guinea
[†] *University of Bayreuth, Faculty of Biology, Postfach 101251, 95440 Bayreuth, Germany*

INTRODUCTION

In Papua New Guinea, large areas with a potential forest cover have been converted for agricultural purposes or were degraded to grasslands and secondary bushland. In 1975, Papua New Guinea's forest cover (including open woodland and scrub) was estimated to exceed 68 percent of the country's land surface (Unesco/UNEP/FAO 1978). This value has continuously decreased since. However, the extent of grassland with a potential for reafforestation is difficult to measure.

The forest cover in densly populated submontane and montane zones has been largely stripped off and accessible lowland forests are continuously being deploited for timber and other wood products.

At an elevation between 1000 and 2500 m, frequently burnt grasslands dominate the landscape in the vicinity of roads and human settlements. In most cases, this fire subclimax vegetation is not productive for gardening, animal husbandry or forestry.

The present paper reports on the findings regarding the nutrient status of a regrowth forest planted in grassland some 50 years ago and compares it with an adjacent *Melinis minutiflora* grassland site. It analyses the source of nutrients which accumulate during succession and evaluates the possibilities of forest regeneration in the area.

NATURAL CONDITIONS OF THE STUDY SITE

The study area is located within the grounds of Wau Ecology Institute at 146°43' E and 7°21' S, 2 km ESE of Wau township. It is situated on the footslopes of Mt. Kaindi at an elevation of 1100 m a.s.l. The area is slightly falling with an inclination of 3° at the study sites.

The underlying bedrock is made up of Mesozoic phyllites and slates of the Owen Stanley metamorphics with some intrusions of volcanic/lahar breccia of the Namie series nearby (Dow *et al.* 1974; Lowenstein 1982).

McAlpine *et al.* (1983) have classified the climate as "humid to perhumid mesothermal with little or no water deficit". The climate diagram of Wau (Figure 1) shows a slight seasonality with a short dry period in June and July.

Data on soil analyses are not published for the study area. Bleeker (1983) mapped Troporthents for the Wau Valley. According to the FAO (1988) classification, haplic Phaeozems cover the mid-montane grasslands up to 2000 m elevation while forests of higher elevation are underlain by humic Cambisols. Louman (unpublished) reported details on physical and chemical properties of the top 30 cm from various plots in the area.

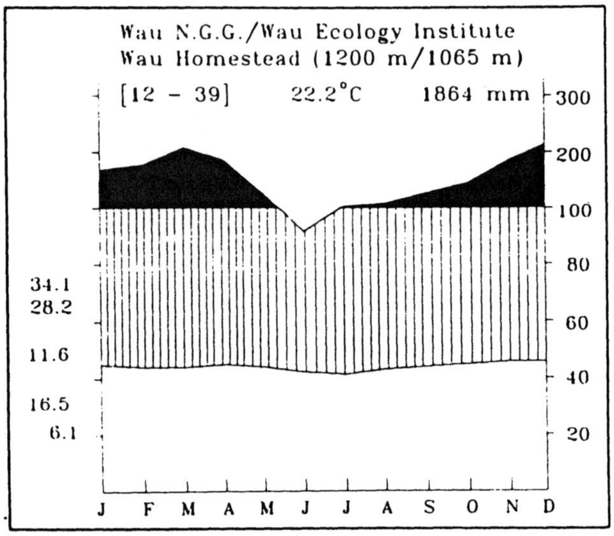

Figure 1. Climate diagram of Wau (rainfall data from Wau N.G.G. [35 years of record] and Wau Ecology Institute [4 years of record]; temperature data from Wau Homestead [12 years of record]). Data cited from McAlpine *et al.* (1983), McAlpine *et al.* (1975) and Wau Ecology Institute (unpublished)

The natural vegetation is a mid-montane evergreen forest with local dominance of one or a few species. Among the common tree species are *Araucaria cunninghamii*, *A. hunsteinii*, *Castanopsis acuminatissima*, *Litsea guppyi*, *Trichospermum plagiophyllum*, *Anthocephalus chinensis*. Large tracts of vegetation have been converted to coffee plantations and various grassland types.

METHODS

Soil sampling, analysis and classification

Soil pits were dug both in the grassland and in the forest and profile description (depth, colour, texture, structure) followed standard methods. Samples of each horizon were taken and dried at 105 °C on the same day. The fraction smaller than 2 mm was used for analysis.

Soil material was analysed by the National Agricultural Chemistry Laboratory using standard methods. Classification followed FAO (1988).

Sampling of plant material

Four plots of 1 m^2 (1 by 1 m) were selected randomly each in frequently burnt grassland and in adjacent regrowth forest. In the grassland plots and in the forest undergrowth, aboveground biomass was collected by cutting the plants at ground level. Sampling of biomass from trees was done by climbing up a rope using the single rope technique (Perry 1978) and collecting the biomass within a metal frame of 1 m^2 which was pulled up the rope.

Since this method would have omitted the stem biomass, an average volume of stem biomass was calculated for a 0.25 ha plot by recording height and diameter in breast height (dbh) of every tree exceeding 2 m height using a volume equation for Southeast Asian forest trees (Bureau of Forestry et al. 1973):

$$V = 5 * 10^{-5} * d^2 * h$$

where V = volume in m^3, d = diameter at breast height, h = height in metres.

The biomass for the plot was calculated as the mean of the specific wood density of six of the common tree species. The respective values for 1 m^2 were added to the branch biomass.

Crude root biomass was obtained by sifting the ground, handcollecting, washing and drying the roots at 80°C. Since fine roots could not be collected with the coarse sieve fine root biomass was determined using 100 ml cylinders and carefully seperating the root material. 2 Cylinders each were taken from within 0 to 5 cm, 5 to 15 and 15 to 30 cm depth, respectively. The material was washed and root tips were counted immediately after collection. The material was then dried and weighed.

RESULTS

Soil

Soil profiles in the grassland and in the regrowth forest plot were both classified as chromic Cambisols containing a high percentage of calcium carbonate. Despite the similar classification, the two soils revealed a number of differences (detailed results will be published elsewhere). The forest soil was more acid (pH~5.0 compared to 5.5 under grassland) and had a lower base saturation (around 70 %) than the grassland soil (100 %). The supply in exchangeable cations, however, was rather similar except for potassium, where the forest soil had a higher capacity. Phosphorus concentration in the grassland profile was high near the surface (37 mg/kg) and decreased drastically with increasing depth. The forest profile showed higher phosphorus concentrations in the lower part of the organic and in the upper part of the mineral horizon. Phosphorus retention, however, was higher in the forest soil (~25 % compared to ~15 % under grassland. The amount of organic substance was higher in the forest soil (about 4 % organic carbon and 0.3 % total nitrogen compared to 2 % organic carbon and 0.15 % total nitrogen in the grassland soil). However, the C/N ratio at the forest soil surface was rather high (17 compared to 13 in the grassland). Texture was similar in the two plots with a slightly higher clay content in the forest soil.

Floristics and Structure

The grassland plots were poor in species (Table 1). They were characterised by the "maygrass" *Melinis minutiflora* Beauv. and a common weed, *Ageratum conyzoides* L. *Piper aduncum* L. was the only shrubby species.

Table 1. Species and their cover/abundance in four grassland plots

Plant species	Plot No.:	1	2	3	4
Melinis minutiflora Beauv.		5	4	5	4
Ageratum conyzoides L.		2	1	2	2
Pueraria phaseoloides (Roxb.) Benth. var. *javanica* (Benth.) Bak.		2	.	.	.
Commelina diffusa Burm. f.		+	.	.	.
Paspalum paniculatum L.		2	.	.	.
Piper aduncum L.		.	3	.	.
Sida rhombifolia L. ssp. *rhombifolia*		.	.	.	1

The regrowth forest was characterised by a stratum of trees around 30 m tall (see forest profile, Figure 2, p. 154), among them *Araucaria cunninghamii* Ait. (Araucariaceae), *Trichospermum pleiostigma* (F. Muell.) Kosterm. (Tiliaceae), *Trema orientalis* (L.) Bl. (Ulmaceae) and *Anthocephalus chinensis* (Lamk.) Rich. (Rubiaceae). The uppermost canopy had a total cover of 70 percent. A number of trees with the potential of reaching the forest canopy had grown between 12 and 25 m tall including *Litsea guppyi* (F. Muell.) F. Muell. ex Forman (Lauraceae), *Aleurites moluccana* (L.) Willd. (Euphorbiaceae) and two species of *Cryptocarya* (Lauraceae). *Euodia cf. radlkoferiana* (Rutaceae) was a common subcanopy species.

Below 10 metres height, *Ficus septica* Burm. f.(Moraceae), *Datura candida* (Pers.) Safford (Solanaceae) and *Piper aduncum* L. (Piperaceae) were the most common species. Depressions with a higher soil water supply were characterised by *Pandanus krauelianus* K. Sch. (Pandanaceae). As Table 2 (p. 155) reveals, the undergrowth varied in cover and was patchily dominated by *Iresine herbstii* Hook. f. (Amaranthaceae), *Rubus moluccanus* (L.) Laut. (Rosaceae), *Elephantopus mollis* H.B.K. (Asteraceae) and *Pteris moluccana* Blume (Adiantaceae, Pteridoideae).

Figure 2. Profile of a regrowth forest in Wau

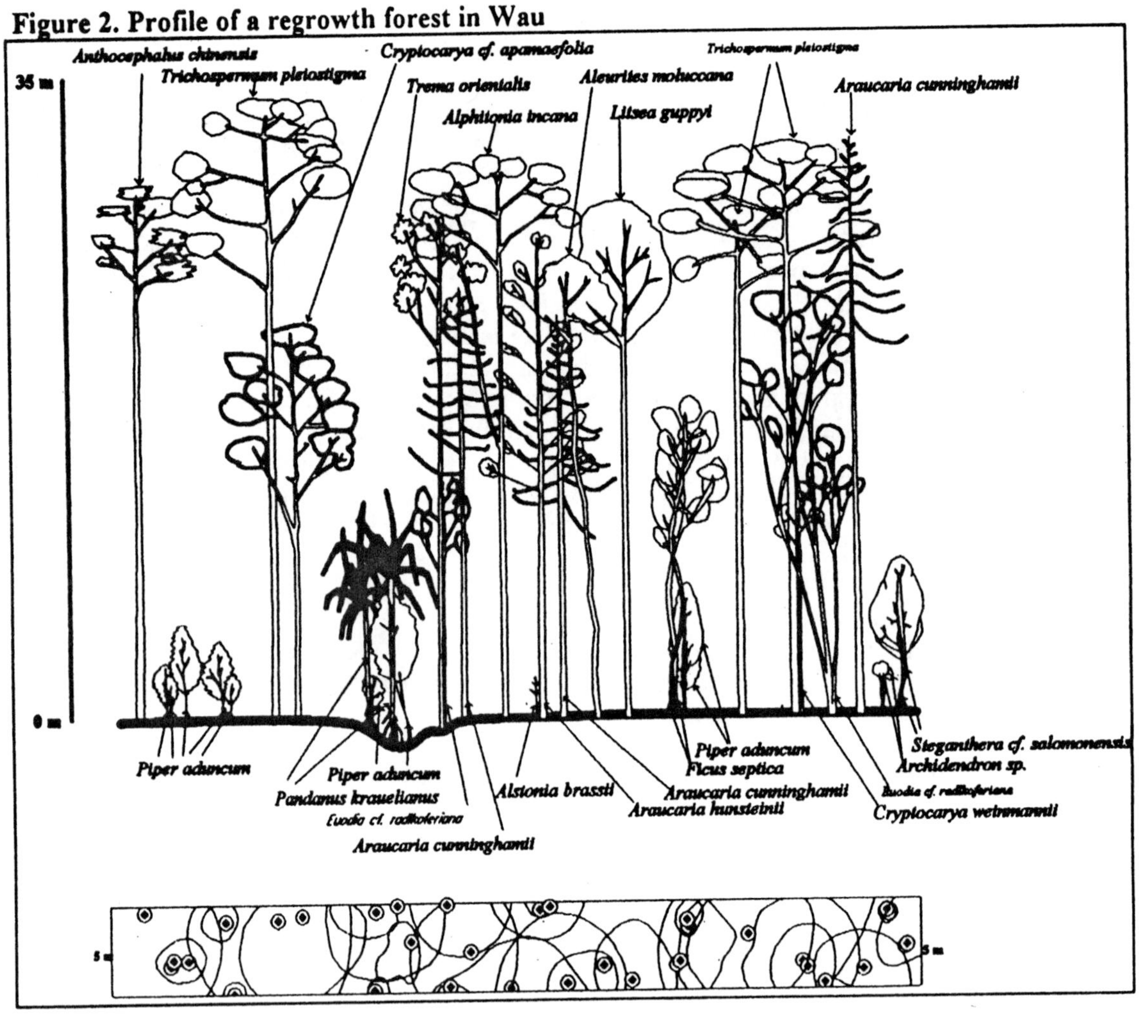

The composition of trees of various diameter classes (Figure 3) showed a depression in the size classes 5-7.5 and 7.5-10 cm diameter. Trees exceeding 50 cm of diameter were not present. The total number of stems with a diameter exceeding 10 cm was 944 per hectare.

Figure 3. Diameter class distribution of trees on 1 hectare of regrowth forest

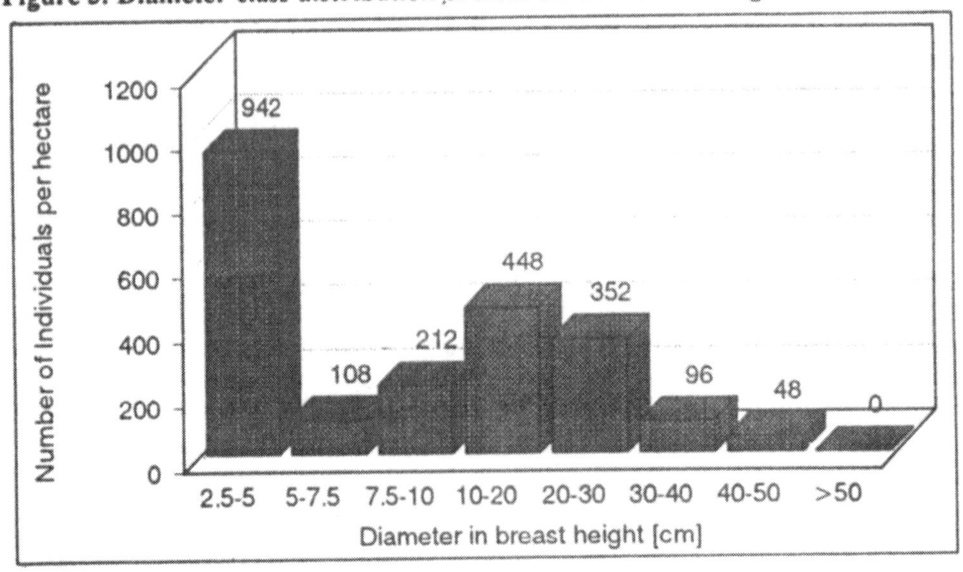

The floristic composition of the study plots within the forest is summarized in Table 2. The life form of lianas is represented only by two fabaceous species (*Mucuna* and *Archidendron*) and their cover is small. Shrubs of open areas and secondary bushland are abundant. The forest floor has a total plant cover (plants < 1 m in height) of 80 percent.

Table 2. Species and their cover/abundance in four forest plots

Plant species \Plot No.	1	2	3	4
Trichospermum pleiostigma (F.Muell.) Kosterm.	.	5	5	5
Acalypha insulana Muell. Arg.	.	1	3	1
Rubus moluccanus (L.) Laut.	2	.	2	2
Elephantopus mollis H.B.K.	.	.	4	+
Pteris moluccana Blume	1	.	1	.
Litsea guppyi (F. Muell.) F. Muell. ex Forman	R	R	.	.
Aleurites moluccana (L.) Willd.	5	.	.	.
Pandanus krauelianus K. Sch.	3	.	.	.
Mucuna novoguineensis Scheff.	1	.	.	.
Datura candida (Pers.) Safford	R	.	.	.
Sphaerostephanus unitus (L.) Holtt.	+	.	.	.
Poaceae gen. indet.	2	.	.	.
Nephrolepis falcata (Cav.) C. Chr.	R	.	.	.
Dysoxylum molle Miq.	.	+	.	.
Iresine herbstii Hook. f.	.	4	.	.
Archidendron sp.	.	R	.	.
Ficus septica Burm. f.	.	.	2	.
Sphaerostephanus deminuens Holtt.	.	.	1	.
Araucaria hunsteinii Ait.	.	.	.	2
Euodia cf. radlkoferiana Laut.	.	.	.	3
Lantana camara L.	.	.	.	1
Piper aduncum L.	.	.	.	2
Cyrtococcum accrescens (Trin.) Stapf	.	.	.	2
Sida rhombifolia L. ssp. *rhombifolia*	.	.	.	+
Centrosema pubescens Benth.	.	.	.	+

The total plant biomass is the grassland was about 2.4 kg per m^2. Plant material in the forest amounted to 34.3 kg per m^2.96 percent of this amount consisted of woody parts. The total wood volume of trees with a diameter exceeding 2.5 cm was 581.72 m^2/ha. The average specific wood weight was 0.4 g per cm^3.

The ratio of aboveground to belowground biomass was less than 2 in the grassland and almost 11 in the forest. Surprisingly, the total amount of litter in the forest plots was rather small (620 g/m^2).

Despite a slightly larger amount of roots in the upper 30 cm of the forest plots, the number of root tips in the grassland was more than three times as high. The fine root biomass below 30 cm depth was neglectable in all plots.

In Table 3 (p. 156), results of the nutrient analyses are summarized. The ratio of nutrients forest *vs.* grassland was high throughout. Nutrient concentrations, however were higher in the grassland biomass. When the entire system including soil nutrients was considered, the total of nitrogen, phosphorus and potassium in the forest exceeded the supplies in the grassland whereas the amounts of calcium and magnesium were equal in both vegetation types.

Table 3. Ratio of macro-nutrients in the forest and in the grassland stored in the respective biomass only and stored in the entire system (biomass and soil)

	Nitrogen	Phosphorus	Potassium	Calcium	Magnesium
Nutrients stored in biomass					
Ratio Forest/Grassland	24.2	20.2	14.5	39.7	28.5
Nutrients stored in biomass and soil					
Ratio Forest/Grassland	2.5	4.6	3.6	1.0	1.1
Nutrient concentration					
Ratio Forest/Grassland	0.52	0.63	0.88	0.32	0.45

The distribution of macronutrients within three compartments (vegetation, litter and soil) from this study did not differ much from various other forest types (Table 4). Obviously, the phosphorus cycling is slower than in most other tropical forests since the percentage of phosphorus in the standing litter was less than 1 percent. Otherwise, nutrient supplies in the soil appeared to be sufficient for a fast regeneration.

Table 4. Distributions of macroelements among vegetation (V), litter (L) and soil (S) compartments in a range of forest ecosystems. Data after (1) Duvigneaud & Denaeyer-De Smet (1970), (2) Westman & Rogers (1977), (3) Tsutsumi (1971), (4, 5) Golley *et al.* (1975), (6) this study. (Data brought to add up to 100 percent, minimum = 1)

No.	Forest type	Soil depth	Nutrient element and storage compartment (%) N			P			K			Ca			Mg		
			V	L	S	V	L	S	V	L	S	V	L	S	V	L	S
1	Temperate	80	10	1	89	5	1	94	1	1	98	1	1	98	2	1	97
2	Eucalypt	200	1	1	98	17	7	76	41	1	58	40	2	58	13	1	86
3	Rainforest	79	8	1	91	71	7	22	31	1	68	44	7	49	59	3	38
4	Rainforest	30				79	8	13	89	1	10	14	1	85	15	1	84
5	Rainforest	25	30	1	69				58	1	41	50	1	49	50	1	49
6	Rainforest	100	14	1	85	76	1	23	68	1	31	12	1	87	30	1	69

Table 5 compares the amounts of biomass, litter and soil organic matter with various other forests. At our study site, the total aboveground and belowground biomass was in the range of other similar forests whereas the amount of leaf biomass was slightly lower and the amount of litter was much lower than in the other studies. The total aboveground biomass of 320 tons/hectare appeared to be high considering the young age of the regrowth forest.

In Table 6, the total amount of macronutrients in the soil is compared with findings from other vegetation types. The total phosphorus content in both the grassland and the forest soils from Wau was significantly lower than in the soils from other areas whereas the amount of other nutrients was rather high. The amounts of calcium were excessivly high and there was a tendency towards an imbalanced magnesium/potassium ratio resulting in an inhibition of potassium uptake in some horizons.

Table 5. Aboveground, belowground and leaf biomass, amounts of litter and soil organic matter in tropical forests of various elevations and status. Data for (1), (3) and (5) cited from (6); (2) Greenland & Kowal (1960), (4) Golley *et al.* (1975), (6) Edwards & Grubb (1977). *Organic matter calculated as twice the organic carbon.

Location (reference)	Altitude (m)	succesional status	Above-ground biomass of living plants ($t \times ha^{-1}$)	Below-ground biomass of living plants ($t \times ha^{-1}$)	Leaf biomass ($t \times ha^{-1}$)	total litter ($t \times ha^{-1}$)	organic matter in the soil ($t \times ha^{-1}$)
Lowland rain forests and semi-decidious forests							
Brazil (1)	50	primary little disturbed	406	67	9.3	24	248 0-100 cm
Ghana (2)	50	secondary ?50 year old	233	54	N.D.	74.3	86 * 0-30 cm
Malaya (3)	100	primary little disturbed	475	N.D.	7.8	52.6	136 * 0-100 cm
Panama (4)	100	secondary <400 years old	316	11	9.3	~19.5	N.D.
Thailand (5)	150	old secondary	365	32	7.8	N.D.	~105 * 0-70 cm
Submontane rain forest							
Papua New Guinea (this study)	1100	secondary ?50 year old	320	23.2	6.8	6.2	493 * 0-100 cm
Montane rain forest							
Papua New Guinea (6)	2450	primary undisturbed	310	40	8.9	18.7	1200 * 0-100 cm

Table 6. Nutrient element contents in the 0-100 cm zone of a range of Australasian soils (Nos. 1-5 and 8 from Australia, data cited from Charley & Richards (1983); **Nos. 6 and 7 from this study**

No.	Annual rainfall (mm)	Parent material	Vegetation	Elemental content ($g \times m^{-3}$) Total N	P	Exchangeable K	Ca	Mg
1	660	Basalt colluvium	Grassland	1490	440	8380	2760	300
2	1070	Basalt	Rainforest	1940	3310	1390	270	1150
3	1575	Shale	Tall eucalypt forest	1370	230	40	80	40
4	1600	Silceous sand	Rainforest	120	20	50	30	90
5	1625	Sandstone	Eucalypt forst	350	110	120	120	70
6	1864	Mesocoic metamorphics	grassland	794	18	103	3215	286
7	1864	Mesocoic metamorphics	submontane forest	1726	23	145	2874	228
8	2135	Basalt colluvium	Alpine woodland	1620	4920	100	30	60

The main source of nutrients for forest regeneration was found to be the soil. The nutrient input from rainfall (bulk precipitation) and other atmospheric inputs was generally low (Table 7, p. 158). The portion of atmospheric nutrient input explaining the difference in total nutrients in the grassland *vs.* the forest system exceeded 10 percent only for calcium.

Table 7. Annual nutrient input by bulk precipitation and contribution of atmospheric nutrients to nutrient accumulation. 1 Data for rainfall input calculated from Edwards (1982); 2 Data for rainfall input calculated from Levett & Mangi (1993); 3 Data for rainfall input from Levett & Mangi (1993) corrected for the marine influence.

	N	P	K	Ca	Mg
1 Nutrient input [g x m^{-2} x a^{-1}]	0.30	0.025	0.34	0.17	0.06
2 Nutrient input [g x m^{-2} x a^{-1}]	0.25	N.D.	0.43	0.84	0.27
3 Nutrient input [g x m^{-2} x a^{-1}]	0.25	N.D.	0.35	0.76	0.07
Maximum input within 50 years [g x m^{-2} x a^{-1}]	15	1.25	15	38	3.5
Portion of atmospheric nutrient input (max. of 50 years) on the amount of nutrients accumulated during 50 years of forest regrowth [%]	0.08	1.6	4.5	81	9.3

DISCUSSION

Forest regeneration appeared to be fast within our study area. The total amount of biomass accumulated within 50 years of regrowth was higher than in most comparable studies (e.g. Greenland & Kowal 1960; Golley *et al.* 1975). The portion of persistent species, however, was low. *Castanopsis acuminatissima* (Bl.) A. DC., being among the dominant species in leftovers of (the mostly disturbed) primary forest, was lacking completely. It was not even found among the saplings and seedlings although the seed source is abundant and regeneration of *Castanopsis* in other disturbed vegetation types is vigourous. Regeneration towards a forest which would be floristically similar to a primary forest is therefore estimated to take a much longer period of time.

The structural analysis of the the regrowth forest revealed an obvious depression in the number of saplings and young trees (trees up to 10 cm diameter). Since larger trees are present in higher numbers, this lack of saplings can be explained only by anthropogenic disturbance. The consequences of a destruction of young trees during the regrowth phase of a forest may become obvious only after several decades. The microclimatic conditions in the undergrowth may - even now - adversely influence the germination and regrowth of persistent tree species and thus have a long-term effect on the future species composition.

The chemistry of soils under annually burnt grassland and under 50 year old regrowth forest did not show much difference. Nutrient input from the forest vegetation to the upper soil horizons may be much lower in the regrowth forest than in a mature forest type since the forest is still in a phase of net-incorporation of nutrients. This is reflected by the small amount of standing litter associated with a limited release of nutrients. The soils are generally well-supplied in all macro-nutrients except of phosphorus (compare Charley & Richards 1983). Phosphorus retention in the forest soil may become an additional severe problem when the amount of organic matter in the top soil increases. Edwards & Grubb (1977) discussed the phenomenon of complexation of phosphorus with organic matter for a montane rain forest in the Chimbu Province of Papua New Guinea.

The results of the soil analyses do not support the observation of a serious nutrient depletion of the topsoil in all grassland areas. A comparison of the nutrient status of primary forest soils with nearby grassland soils would be desirable in this respect. Our data suggests that great care should be taken not to generalize results from a restricted number of studies. The limited size of our study area whose soils are partly derived from volcanic outcrops and who are thus likely to be richer in nutrients than average have to be taken into consideration. Furthermore, the small number of sample plots, especially in the forest, and the limitations of biomass sampling from the tree crowns (sampling closer to the trunk than average) impose a restriction to the data set. However, the biomass data obtained with our non-destructive approach compares well with other studies. Limitations of root biomass sampling in tropical forests have been widely discussed (e.g. Jeník 1971) and we are aware of the restrictions associated with our methods (missing out of large vertical sinkers and roots below 1 m depth). Despite this underestimation of the actual root biomass the values obtained from our study are rather high.

Some uncertainty is associated with the informations on the history of the plot. Floristically, the regrowth character of the forest plots is obvious. However, the time and type of reafforestation is unknown. It can be concluded from the diameter and height composition of the forest that regrowth was undisturbed for an initial 30 to 40 years. Disturbance of the plot has obviously begun only recently and if it continues in the same way the regrowth forest will eventually be altered to secondary bushland. Presently, the canopy is made up of short-lived, light wooded trees occurring in secondary forests and open areas. The plot therefore depends on the undisturbed growth of potential canopy trees to fill out the future treefall gaps.

Conclusively, the area seems to be generally promising for reafforestation with a mixed set of tree species. In the present situation, enrichment plantings with persistent tree species would be recommended to fill out the gaps created be anthropogenic influence. However, such measure can only be successful if the felling of trees and the extraction of dead wood can be stopped until the age composition of the stand is balanced and the percentage of persistent trees has increased.

ACKNOWLEDGEMENTS

We are indebted to Wau Ecology Institute for supporting all aspects of the study. Yves Basset allowed us the use of his climbing gear. Fred Grieshaber made valuable comments on the soil classification. Paul Katik assisted in the identification of the plant material. Field assistance was provided by Andrew Kumbol. The study was partiall funded by the A.F.W. Schimper scholarship of the H. & E. Walter foundation.

REFERENCES

Bureau of Forestry, U.P. College of Forestry, U.S.A.I.D. (1973). *Regional volume equations and tables for Philippine timber species.* Manila.

Bleeker, P. (1983). *Soils of Papua New Guinea.* CSIRO and Australian National University Press, 352 pp.

Charley, J.J. & Richards, B.N. (1983). Nutrient allocation in plant communities. *Physiological Plant Ecology IV. Ecosystem Processes: Mineral Cycling, Productivity and Man's Influence.* (Ed. by O.L. Lange, P.S. Nobel, C.B. Osmond & H. Ziegler), pp. 5-45, Springer Verlag Berlin, Heidelberg, New York.

Dow, D.B., Smit, J.A.J. & Page, R.W. (1974). *1:250 000 Geological Series - Explanatory Notes: Wau, Papua New Guinea, Sheet SB/55-14 International Index.* Australian Government Publishing Service Canberra.

Duvigneaud, P. & Denaeyer-De Smet, S. (1970). Biological cycling of minerals in temperate deciduous forests. *Ecological Studies, vol. I. Analysis of temperate forest ecosystems.* (Ed. by D.E. Reichle), Springer Verlag, Berlin Heidelberg, New York.

Edwards, P.J. (1982). Studies of mineral cycling in a montane rain forest in New Guinea. V. Rates of cycling in throughfall and litter fall. *J. Ecol.* **70,** 807-827.

Edwards, P.J. & Grubb, P.J. (1977). Studies of mineral cycling in a montane rain forest in New Guinea. I. The distribution of organic matter in the vegetation and soil. *J. Ecol.* **65,** 943-969.

FAO (1988). *FAO-Unesco Soil Map of the World, revised legend.*- World Soil Resources Report 60, FAO, Rome.

Golley F.B., McGinnis, J.T., Clements, R.G., Child, G.I. & Duever, M.J. (1975). *Mineral cycling in a tropical moist forest ecosystem.* University of Georgia Press, Athens, Georgia.

Greenland, D.J. & Kowal, J.M.L. (1960). Nutrient content of the moist tropical forest of Ghana. *Plant and Soil* **12 (2),** 154-174.

Jeník, J. (1971). Root structure and underground biomass in equatorial forests. La productivité des écosystèmes forestiers (Ed. by P. Duvigneaud), pp. 323-331, Unesco, Paris.

Levett, M.P. & Mangi, V. (1993). Bulk precipitation quality in the Port Moresby area of Papua New Guinea. *Science in New Guinea* **19(1),** 29-35.

Louman, B.T. (unpublished). Structural and floristic composition of a *Lithocarpus* fallow in Papua New Guinea. Unpublished MSc-thesis, Wageningen No. 86/15, 1986.

Lowenstein, P.L. (1982). *Economic geology of the Morobe Goldfield.* Geological Survey of Papua New Guinea Memoir 9, Libra Press Hong Kong, 2 vols.

McAlpine, J.R., Keig, G., Falls, R. (1983). *Climate of Papua New Guinea.* Commonwealth Scientific and Industrial Research Organization (Australia) 200 pp.

Mc Alpine, J.R., Keig, G., Short, K. (1975). *Climatic tables for Papua New Guinea.* Division of Land Use Research Technical Paper No. 37, Commonwealth Scientific and Industrial Research Organisation, Australia; CSIRO, Melbourne, 177 pp.

Keig, G., Fleming, P.M., McAlpine, J.R. (1979). Evaporation in Papua New Guinea. *J. Trop. Geogr.* **48,** 19-30.

Perry, D.R. (1978). A method of access into the crowns of emergent and canopy trees. *Biotropica* **10,** 155-157.

Tsutsumi, T. (1971). Accumulation and circulation of elements in forest ecosystems. *Productivity of forest ecosystems.* (Ed. by P. Duvigneaud), Ecology and conservation 4, Unesco, Paris.

Unesco/UNEP/FAO (1978). *Tropical forest ecosystems.* Natural resources research XIV, Unesco Paris, 683 pp.

Wau Ecology Institute (unpublished). Weather readings for 1990 to 1993. Wau Ecology Institute.

Westman, W.E., Rogers, R.W. (1977). Nutrient stocks in a subtropical eucalypt forest, North Stradbroke Island. *Aust. J. Ecol.* **2,** 447-460.

Proceedings of the Biological Society of New Guinea
R. Höft (ed.)

RATES OF LITTER DECOMPOSITION IN DIFFERENT HABITATS IN BULOLO AREA, MOROBE PROVINCE

Balun Lawong, Tolusina Pouli & Galus Wasukware Anton
Papua New Guinea Forestry College, P.O.Box 92, Bulolo, Morobe Province, Papua New Guinea.

ABSTRACT

In a hypothetical steady state ecosystem, nutrients taken up by plants are balanced by those returned to the soil with dead plant parts (litterfall) or leached from the plants by rain (throughfall). This study investigated the rates of litter decomposition between different species (*Castanopsis accuminatissima* and *Eucalyptus territicornis*) and habitat types (*Castanopsis* forest, mixed forest, and grassland) in the mid-montane zone.

The rates of litter decomposition varied significantly ($P<0.001$) between different species. The two species showed different rates of weight loss (g) in similar habitat types. The rates of litter decomposition varied significantly ($P<0.001$) between different habitat types. The rate of litter decomposition was highest in the *Castanopsis* forest type, followed by the mixed forest.

Rates of litter decomposition show some correlation with local temperatures, light, soils and humidity prevailing in the different habitat types.

The significant difference ($P<0.001$) in rates of litter decomposition between different habitat types shows that the rate of decomposition is not entirely dependent on altitude.

INTRODUCTION

Litter Decomposition is the loss of organic matter or the physical breakdown of plant or animal matter and the subsequent release of essential growth elements (nutrients) into the soil. Polunin (1984) showed that the amount of organic matter in the litter and humus layer accounted for 84% of the observed variance in the annual nitrogen mineralization. Litter decomposition occurs both in the above and below ground. The below-ground litter decomposition refers to the death and decay of fine roots, mycorrhizae and associated microflora. This study concentrated on the above-ground leaf litter decomposition.

Forest litter is classified into two types: (a) coarse litter: this includes fallen branches and boles. Due to the heterogeneity in space and time this is difficult to assay and (b) fine litter: this includes small twigs, leaves, flowers and fruits. In the humid tropics, floral parts may begin the decay process while still on the plant, and decomposition with the exception of certain highly resistant wood, proceeds rapidly on the soil surface. The rate at which litter falls and decays can be determined by collecting the falling organic material in suitable traps and by isolating the fallen material on the ground and re-examining it at certain intervals.

Substantial proportions of organic substances from plants or animals eventually remain on the soil surface for only a few hours or days if they are readily decomposable. However, retention can be as long as several decades if they decay slowly (Jordon, 1989). The rate of breakdown of litter varies considerably between seasons and sites.

A direct measure of the rate of decomposition is impossible.An indirect measure from dry weight loss, howevwe can give a crude rate of nutrients deposition. Decomposition is roughly assessed as a weight loss from coarse mesh bags, and usually leaching. Weight loss from the litter typically exhibits a high initial rate which subsequently declines.

The decomposition of leaves, stems and other debris is habitat-specific. Jordan (1989) showed that different forest types and localities have their own rate of litter decomposition. The rate of decomposition on species level is also largely influenced by plant internal factors.

· The rate of litter decomposition has been used as a crude measure of rate of nutrient release into the soil and the availability of nutrients in the nutrient cycle. Many studies have been done on litter fall and decomposition (Turvey, 1974; Edwards & Grubb, 1977; Edwards, 1977; Kato et al, 1978; Edwards, 1982). There are only two known related studies done in Papua New Guinea to date (welche ?). These two studies investigated the rate of litter decomposition in a single forest type. This study differ from the two in that it envisage to study the rate of litter decomposition between different habitat types and species.

STUDY SITE DESCRIPTION

The study was conducted in the Bulolo Valley, Morobe Province, Papua New Guinea (07° 10-15'S, 146° 35-40'E), within an altitudinal range of 600-1400 m (Fig. 1).

The area is subjected to a dry perhumid climate with an annual rainfall of 1512 mm and a mean annual temperature of about 24.3° C (Bulolo station 27 years record, 1933-1990). Soils from the four different sites studied are generally lateritic with a pH range of 5-6. The soils had humus layers ranging in thickness from 0-1 cm in the grassland to 1-4 cm in the forest types.

The study sites consisted of three distinct vegetation types. The vegetation types varied in floristic composition. They comprised *Castanopsis* forest, mixed forest and grassland. Except for the grassland the floors were covered with seedlings and saplings, which were derived largely from the dominant species. The life forms in the ground vegetation ranged from herbs to shrubs and understorey trees. The grassland was dominated by *Arudinella* spec. and *Imperata cylindrica*.

METHODS

Leaf litter samples were collected from two plant species; *Castanopsis accuminatissima* and the *Eucalyptus territicornis* and were dried in the oven for one week. Dried litter from each species was then measured into 20 replicates, concealed in sarlon cloth bags (*ca*. 20 x 15 x 2 cm) and labelled with an aluminium tag. The initial dry weight readings were taken from the electric balance. Five replicates of the samples were then exposed to three different habitat types and to room temperature. These habitat types were the *Castanopsis* forest, mixed forest and Savanna grassland. Samples were recollected from the field and taken back to the laboratory for drying and reweighing on monthly basis for 6 months over a two year period.

Fine litterfall

The data from fine litter fall was used for calculation of the decay factor (k). Five litter traps (1 x 1m) with conical-shaped depression frames clothed with sarlon were placed at approximately two metres height above ground at 40 m intervals in the *Castanopsis* and mixed forest. The litter samples were collected on a monthly basis over a period of six months for two years. The litter samples were dried at 50 °C to a constant dry weight point and the weights were recorded.

RESULTS

The results obtained are presented in the Tables 1, 2 and 3 and Figures 1 and 2. Table 1 shows the cumulative mean weight loss (g) within six (6) months in different times at different sites

Table 1. Cumulative mean weight loss of two types of litter in four habitats

	Grassland		Castanopsis		Mixed Forest		Control	
	Cas.	Euc.	Cas.	Euc.	Cas.	Euc.	Cas.	Euc.
T0	0	0	0	0	0	0	0	0
T1	9.82	9.40	14.31	13.44	11.23	15.75	1.52	0.49
T2	11.68	11.26	22.02	15.59	13.54	19.33	3.37	1.24
T3	13.83	14.32	25.66	20.30	17.63	23.27	3.69	1.45
T4	15.57	17.02	29.61	23.71	21.50	26.40	3.93	1.61
T5	16.44	17.61	31.56	25.23	22.92	27.53	4.02	1.67
T6	16.72	17.79	32.02	25.59	23.28	27.74	4.08	1.71

Cas = Castanopsis leaf litter, Euc = Eucalyptus litter, T0, T1, T2, T3 are times (months); T0=Initial time, T1 = 1st month, T2 = 2nd month, T3 = 3rd month.

The same results are presented in Figures 1 and 2.

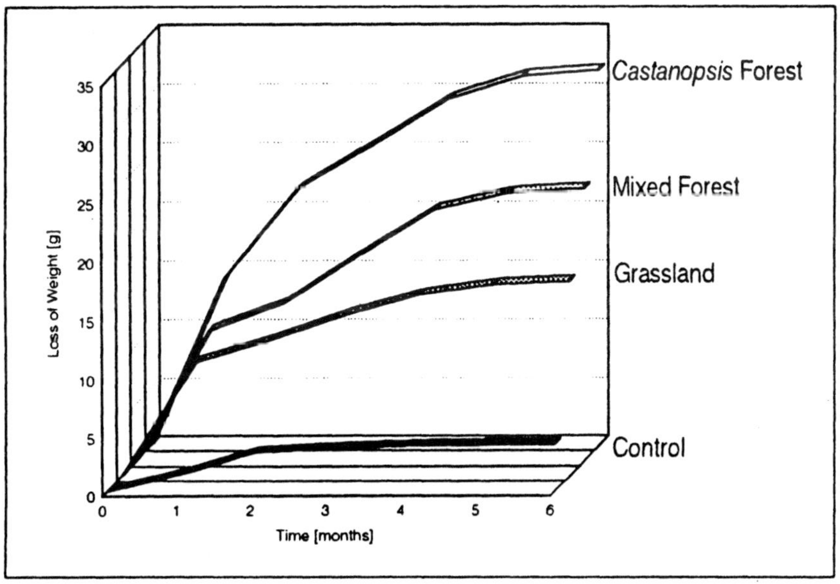

Figure 1. Decomposition of *Castanopsis* litter in four habitats

The result on the ANOVA (Table 2) shows that there is a significant variation in the rate of litter decomposition between different species and habitat types.

Table 2 Nested ANOVA showing F-values

Source	SS	dF	M.S	Fs	
Among Sites	2804.7	3	934.9	9.9	P<0.001
Among Species	4484.9	1	4484.9	47.5	P<0.001
Between Time	5959.1	24	248.3	2.6	
Measurements	12379.0	131	94.5		
Total	25627.7	159			

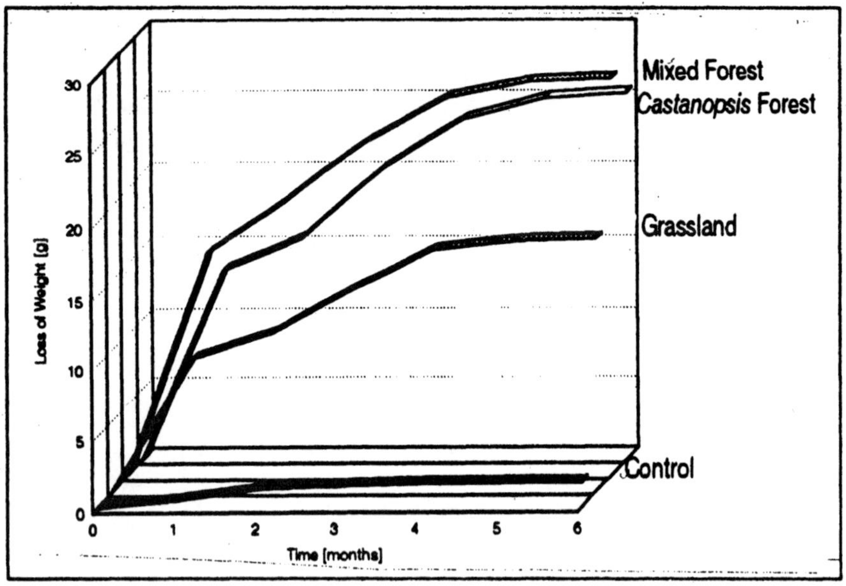

Figure 2. Decomposition of *Eucalyptus* litter in four habitats

Table 3. Comparison of litterfall (tonnes ha^{-1} year^{-1}). Data from the present study as compared to previous related studies

Forest types and place		(a) Fine litterfall (t ha^{-1} year^{-1})	(b) Forest floor (t ha^{-1})	(c) Decay factor, k ((a)/(b))
Lowland evergreen rain forest				
Mulu: ridge (Anderson & Swift, 1983)		7.7	5.9	1.3
Lower montane rain forest				
New Guinea, 4 close sites (Edwards, 1977)		6.2-6.6	4.2-6.6	1.0-1.5
New Guinea, Bulolo (Enright, 1978)		7.7	7.7	1.0
Present Study:	Mixed forest-type	6.3	5.6	1.1
	Castanopsis forest-type	4.2	4.2	1.0

DISCUSSION

The results of this study (Table 2) showed that rates of fine litter decomposition varied significantly ($P<0.001$) between species and habitat types. The rate of decomposition was highest in the *Castanopsis* forest, followed by mixed forest and grassland respectively. Rates of litter decomposition between the species showed *Castanopsis accuminatissima* litter to have decomposed faster than *Ecalyptus territicornis* in *Castanopsis* forest. In the mixed forest, it was the other way round. In the grassland habitat *Eucalyptus territcornis* litter decomposed faster than *Castanopsis accuminatissima* (Figures 1 and 2).

The rate of fine litter decomposition was highest in the first two months and gradually decreased after the third month. The results suggest that in the tropics under closed canopy fine litter rapidly decomposes (within 2-3 months). Obviously the plant species in the tropics must acquire certain strategies including increasing efficiency in nutrient uptake from the soil and retain much of the nutrients in the above ground biomass. Ombrotrophs (trees which have the ability to absorb nutrients directly from the surrounding atmosphere) employ a different strategy.

The release of nutrients from the decaying organic matter into the soil is a critical step in the ecosystem function. If nutrients are released too fast, they can be lost through leaching from the soil or volatilization. Sincec plants take up only a certain amount of nutrients at a certain period of time, the rapid release of nutrients into the soil as indicated in this study is not to the advantage of the standing plant community. In contrast, if decomposition is too slow, insufficient nutrients are made available to plants, with the result that tree growth can be inhibited.

The environmental conditions which favour high rates of nutrient recycling and biomass production also create the potential for high rates of leaching and weathering of tropical soils.

Moisture and Temperature Interactions

These are important factors responsible for differences in nutrient cycling between forest types and localities. Polunin (1984) pointed out that high temperatures occur year-round, and biological processes proceed rapidly if moisture is not limiting. Because of high temperatures year round, nutrient cycling can take place throughout the year in the humid tropics, and annual nutrient cycling can be higher than in regions where cold and drought interrupt these processes.

The values for decomposition are in general similar to those given by decay factor(k) (Table 3). Figures 1 and 2 show that the rate of decomposition in grassland is relatively slow compared to that under *Castanopsis* and mixed forest types. The higher rates of decomposition in the two forest types are obviously due to the higher percentage of relative humidity under the forest canopy. It may be concluded that nutrients in the fine litter under forest canopy are released faster than in the open grassland. Every year, fine litter add to the forest floor an average of 4 to 6 tons x ha^{-1} x $year^{-1}$ (Table 3) hence contributing a considerable proportion of nutrients to the soil nutrient pool.

The Leaf Litter Composition

The rate of litter decomposition at species level is also influenced by species specific internal factors. Some species have thicker and tougher leaves which take time to break down, while others are thinner and softer making it easier for the decomposers. Leaves of different species vary in their chemical contents. The leaves which have chemicals that are tasty to decomposers would be faster colonized by decomposing organisms. In the *Castanopsis* forest type, *Castanopsis accuminatissima* litter decomposed faster than *Eucalyptus territicornis* litter (compare Figures 1 and 2). This might be explained in terms of decomposer species abundance.

ACKNOWLEDGEMENTS

We wish to thank the following people for making a successful completion of this project possible; Mr Tommy Nahuet (Principal Papua New Guinea Forestry College) for allowing us to use the College facilities, Mr Wanaki (PTB Driver) for driving us into the study sites, Mr Gwaimbo Tuk (Assistant Technician-Botany section) for technical assistance and Miriam Todiki for assisting in typing this paper.

REFERENCES

Anderson, J.M. & Swift, M.J. (1983). Decomposition in tropcial rain forests. *Tropical rain forest ecology and management* (Ed. by S. L. Sutton, T.C. Whitmore & A.C. Chadwick), Blackwell, Oxford.

Edwards, P.J. (1977). Studies of mineral cycling in a montane rain forest in New Guinea. II The produciton and disappearance of litter. *J. Ecol.* **65,** 971-92.

Edwards, P.J. (1982). Studies of mineal cycling in a montane rain forest in New Guinea. V Rates of cycling in throughfall and litterfall. *J. Ecol.* **70,** 807-27.

Enright, N.H. (1978). The effect of logging on the regeneration nutrient budget of Araucaria cunninghamii dominated tropical rain forest in Papua New Guinea. *Malaysian Forester* **41**, 303-318.

Edwards, P.J. & Grubb, P.J. (1977). Studies of mineral cycling in a montane rain forest in New Guinea. I The distribution of organic matter in the vegetation and soil. *J. Ecol.* **64,** 943-60.

Jordon, C.F. (1989). *An Amazonian Forest. The Structure and Function of a Nutrient Stressed Ecosystem and the Impact of Slash and Burn Agriculture*. Man and Biosphere Series 2. UNESCO and Parthenon, Carnforth, U.K.

Kato, R., Tadaki, Y., & Ogawa, H. (1978). Plant biomass and growth increment studies in the Pasoh forest. *Malay. Nat. J.* **30,** 211-24.

Polunin, C.V.N. (1984). The Decomposition of Emergent Macrophytes in Fresh Water. *Advances in Ecological Research* **14,** 115-216.

Turvey, N.D. (1974). Water in the nutrient cycle of a Papuan rain forest. *Nature, Lond.* **251,** 414-415.

Session VIII: Conservation: Concepts, Problems, Needs

Proceedings of the Biological Society of New Guinea
R. Höft (ed.)

RESEEDING OF SELECTIVELY FELLED HARDWOOD FORESTS FROM BUFFER ZONES AT OPEN BAY TIMBER AREA, PAPUA NEW GUINEA

Robert Kiapranis & Kipiro Damas

Papua New Guinea Forest Research Institute, P. O. Box 314, Lae, Papua New Guinea

SUMMARY

The aim of this study was to investigate seed dispersal of timber tree species from established river buffer zone forests into selectively logged forests. The project site selected for this study was at Open Bay Timber area in East New Britain Province. Prior to the establishment of this study site, a general survey was carried out on the general status of established buffer zone forests in other timber areas particularly in New Ireland and the Morobe Provinces. It was observed during this survey that the status of the established river buffer zones forests in several logging areas were inconsistent with required regulations of Papua New Guinea. Most river buffer zones observed had less than the required 50 m width, and were often either selectively felled at low intensity or completely destroyed by the local people soon after establishment.

The timber tree species located within the buffer zone forests studied consisted of Potential Seed Sources (74 %) and Seed Sources or Parent Trees (26 %). The common timber species included *Pometia pinnata* (Sapindaceae), *Dysoxylum gaudichaudianum* (Meliaceae), *Celtis latifolia* (Ulmaceae), *Garcinia* sp. (Guttiferae), *Protium* sp. (Burseraceae) and *Canarium indicum* (Buseraceae).

The comparison between the timber species extracted from the areas studied and the species composition of the buffer zones forests showed that 51 percent of the species extracted were not represented within the buffer zone forests. *Calophyllum* sp. which made up 22 percent of the total logged species was represented by only 2 individuals in the buffer zone forests.

Soil seed bank and seed rain species composition from the plots established in two buffer zone forests showed a low correlation with the standing vegetation. The seed rain species composition consisted of 317 seedlings belonging to 48 species (10 seedlings were of timber tree species) of 36 families.

It was observed that seed dispersal into the selectively logged forests peaked at two different times during the study period. The first occurred at the peak period of the rainy season and the second at the peak of the dry season. Most of the timber tree species were recorded during the wet season while the non-timber tree species were dominant during the dry period.

The low recruitment of timber species in the seed rain indicate that seed dispersal from buffer zone forests into logged-over forests is minimal and may depend on the availability of each species or its parent trees. In order to maintain as many timber tree species as possible within the logged forests, the buffer zone areas should be increased so that more parent trees can be preserved.

The results of this study indicate that seed dispersal from timber trees within the buffer zones into logged over forests is lower than assumed. Therefore the buffer zones can not be relied upon for regeneration of logged over forests. Further longer study periods are required to confirm these results.

ACKNOWLEDGEMENTS

We are indebted to the management of the Open Bay Timber Company for allowing us to carry out this study in their project area and for their cooperation and assistance over the project period. We would also like to thank Mr. F. Asok for assisting us in the field and monitoring of samples, Mr. O. Gideon and other colleagues for their assistance and valuable comments. This study was funded by WWF under its Biodiversity Support Program (BSP) Grant # 7533.

RELEVANT LITERATURE

Augspurger, C.K. (1986). Morphology and dispersal potential of wind dispersed diaspores of Neotropical trees. *Am. J. Bot.* **73(3),** 353-363.

Beehler, B. (1983). Frugivory and polygamy in birds of Paradise. *The AUK* **100(1),** 1-12.

Cruz, A. (1981). Bird activity and seed dispersal of a montane forest tree (*Dunalia arborescens*) in Jamica. Reproductive Botany, 34-44.

Diamond, J.M., Raga, M.N., Wiakabu, J., Maru, T. & Feni, S. (undated). Fruit consumption and seed dispersal by New Guinea birds. Report prepared for the Department of Forest, Papua New Guinea.

Driscoll, P.V. (1984). *The effects of logging on bird populations in lowland New Guinea rainforest.* Ph.D. Thesis, University of Queensland, Australia.

Enright, N. (1985). Existence of soil seed bank under rainforest in New Guinea. *Aust. J. Ecol.* **10,** 67-77.

Fleming, T.H & Williams, C.F. (1990). Phenology, seed dispersal and recruitment in *Ceriop peltata* (Moraceac) in Costa Rican tropical rain forest. *J. Trop. Ecol.* **6,** 163-178

Foster, M.S. & McDiamond, R.W. (1983). Nutrional value of the of aril of *Trichilia cuneata*, a bird dispersal fruit. *Biotropica* **15(1),** 26-31.

Frith, H.J., Crome, H.J. & Wolfe, T.O. (1976). Food of fruit-pigeons in New Guinea. *The Emu* **76(2),** 49-58.

Pratt, T.K. & Stiles, E.W. (1985). The influence of fruit size and structure on the composition of frugivorous assemblages in New Guinea. *Biotropica* **17(4),** 314-321.

Saulei, S.M. & Swaine, M.D. (1988). Rain forest seed dynamics during succession at Gogol Valley, Papua New Guinea. *J. Ecol.* **76,** 1133-1152.

Snow, D.W. (1981). Tropical frugivorous birds and their food plants: A world survey. *Biotropica* **13(1),** 1-14.

Whitmore, T.C. (1986). *Tropical rain forests of the Far East.* Second edition. ELBS, Oxford University Press, UK.

Proceedings of the Biological Society of New Guinea
R. Höft (ed.)

TOWARDS CONSERVATION AND INDUSTRY DEVELOPMENT. ORCHID RESEARCH IN PAPUA NEW GUINEA

N.H.S.Howcroft

Papua New Guinea Forest Research Institute, Genetic Resources Research and Development Section, P.O.Box 314, Lae, Papua New Guinea

ABSTRACT

A brief overview, for the need and role of orchid conservation and research to help develop and manage the genetic resources of the indigeneous orchids in Papua New Guinea, and for the rest of Papuasia, is provided in the paper. Economic incentive is seen as the main means of making orchid conservation successful in Papuasia. Problems and advantages of imposing conservation measures are briefly discussed. The progress of Orchid research in Papua New Guinea and its future direction is outlined. Comments are provided on the effectiveness of the 1990 ban on the export wild orchids.

INTRODUCTION

The Forest Research Institute (FRI) and the Foundation For the People of The South Pacific (FSP/Papua New Guinea) are cooperating to establish a flora conservation program (FCP) for Papua New Guinea. The project aims to assist local communities in the conservation of indigeneous flora with horticultural potential and to develop an industry based on flower and plant production, The project places some emphasis on the orchids for which new Guinea is so well known for. It is comprised of several components. These are research, development and marketing.

In this paper the need for, and the progress of, the research component is outlined and discussed.

NEED FOR RESEARCH AND CONSERVATION

Conservation for Whom?

A colleague in a responsible government department was heard and quoted to say 'Conservation, for whom?' In view of the need for economic development at village level to improve rural living standards, this is a reasonable question, when conservation measures may prevent landowners from earning an income from resources on their own land.

The private sector in Papua New Guinea is keen to develop their resources on a sustainable basis, but there are also those members of the sector that are just as keen to exploit their resources without a thought about the future and sustainability.

To be able to develop a sustainable resource there is a need to practice conservation so as to be able to manage the resource in perpetuity. Since the resources belong to the people, who will seek ways and means to earn a living from those resources, conservation will only be possible and acceptable if there is an economic incentive.

Is Orchid Conservation Needed?

We know from world wide experences that if an orchid species is popular that people will exploit it until it can no longer be found in its native habitat. Equally well known is that many orchids cannot survive if their habitat destroyed. Both situations exist in New Guinea and how much has been lost will never be known.

At present very little is known about the conservation needs of the orchids of Papuasia. This is due to the lack of adequate information on their distribution, habitats and biology. Cribb (1991) considers the major threat to the orchids of the South west Pacific to be massive ignorance of the floras. This too may apply to Papuasian orchid flora. Since Papuasian orchids have been collected by botanists and by orchid collectors for commerial purposes or for private collections, for over a century, this lack of information may be surprising to some people.

However, when one looks at the collection locality information of herbarium material, as well as published information, of any of the well known species that have been collected in the past, it can be seen that many of these appear to occur in only a few disjunct areas, whereas in reality the species is likely to have a far greater geographic range than the records indicate (Examples: Distribution of *Spathoglottis plicata* ssp. *nov. puberula*, Maps 1A and 1B and of *Spathoglottis parviflora* and related species, Maps 2A and 2B).

Map 1A. Distributions of *Spathoglottis papuana*, *S. plicata*, *S. plicata* proposed subspecies *bismarckii*, *S. humilis*, *S. neobrittanica*, *S. puberula* and *S. schlechteri* before a thorough survey was made.

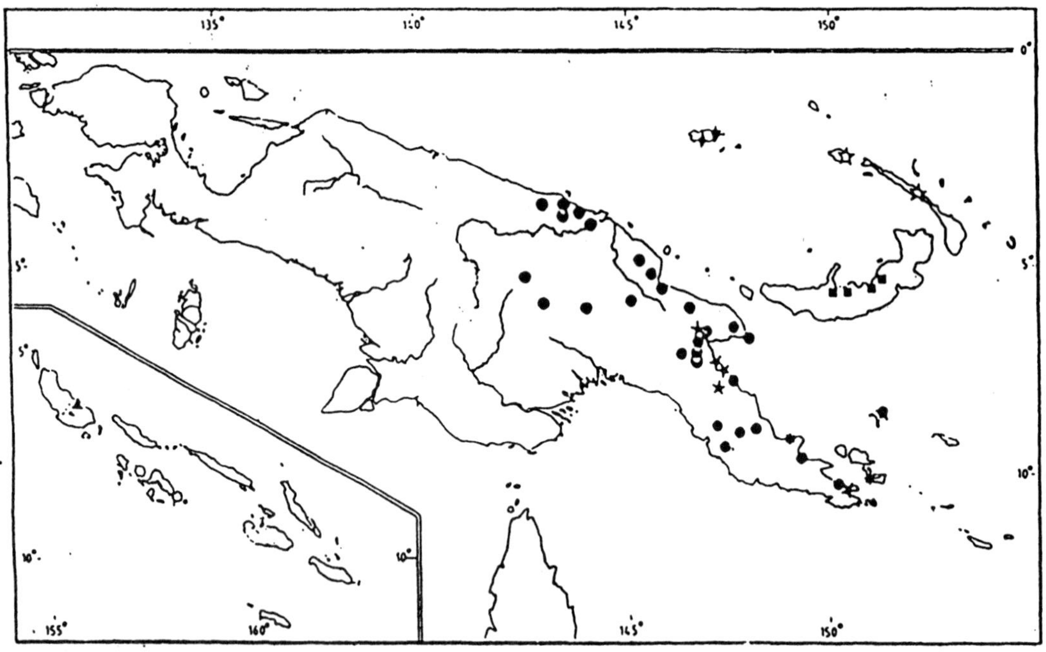

Distributions of S. papuana (✱); S. plicata (▲); proposed subspecies: bismarckii (♦); humilis (★); neo-britannica (■); puberula (●); schlechteri (☆).

Map 1B. Distributions of *Spathoglottis papuana*, *S. plicata*, *S. plicata* proposed subspecies *bismarckii*, *S. humilis*, *S. neobrittanica*, *S. puberula* and *S. schlechteri* after a survey.

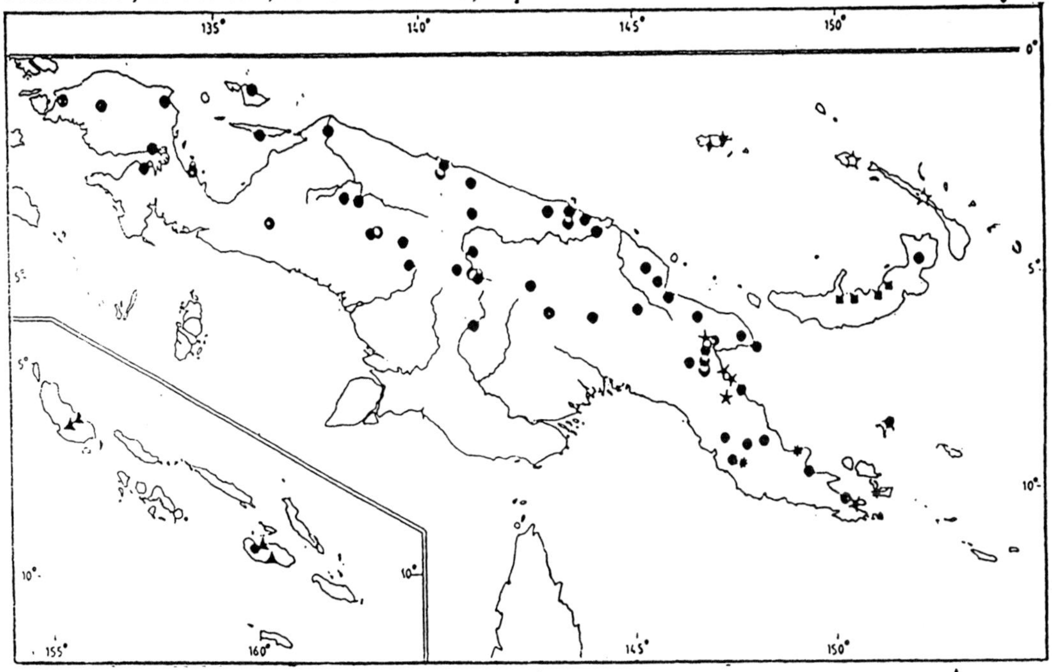

Distributions of S. papuana (✱); S. plicata (▲); proposed subspecies: bismarckii (♦); humilis (★); neo-britannica (■); puberula (●); schlechteri (☆).

Map 2A. Distributions of *Spathoglottis doctersii*, *S. doctersii* var. *emarginata*, *S. parviflora* proposed subspecies *lanceobracteata*, hybrid X *mendiensis* before a thorough survey was made.

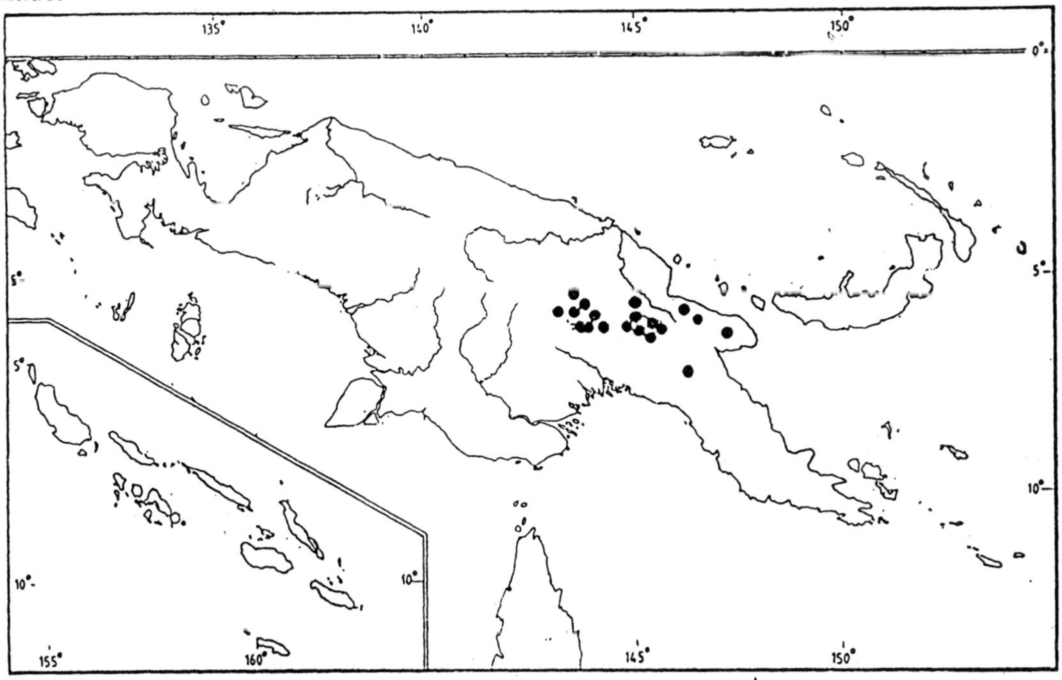

Distributions of S. doctersii (★); S. doctersii var emarginata (★); S. parviflora (●); proposed subspecies: lanceobracteata (○). hybrid X mendiensis (✱)

Map 2A. Distributions of *Spathoglottis doctersii*, *S. doctersii* var. *emarginata*, *S. parviflora* proposed subspecies *lanceobracteata*, hybrid X *mendiensis* after the survey.

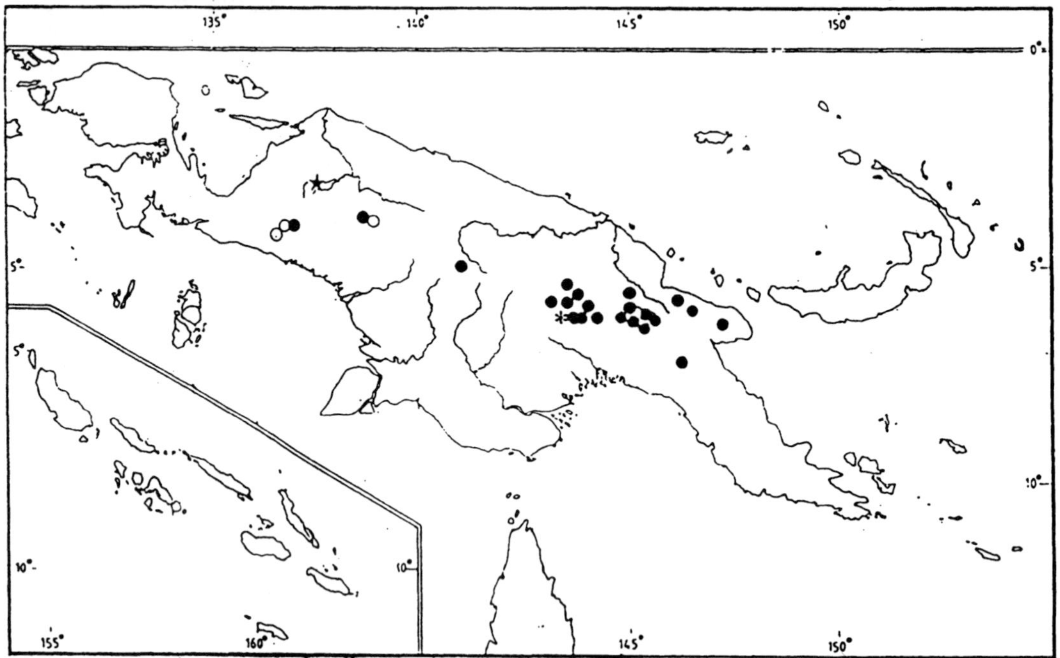

Distributions of S. doctersii (★); S. doctersii var. emarginata (★); S. parviflora (●); proposed subspecies: lanceobracteata (○). hybrid X mendiensis (✳).

On the other hand, the small amount of information that is available might also indicate that the species is uncommon. The important aspect of all of this is that we just do not know, for many of the species, their full geographic range, their habitat requirements, the pollinator behaviour and the breeding structure of the species populations or their conservation needs.

There is then an obvious need for conservation and research with the Papuasian orchids. To establish these the geographic range of each species should be mapped, the habitat association and requirements of each species should be studied and recorded, and to identify those under threat and their conservation needs.

Taxonomic Research.

The diversity and richness of New Guinea orchids is such that new genera and species continue to be discovered to the present. However, many orchid genera and species in Papuasia also need to be the subject of further taxonomic research to solve identification problems that continue to cause confusion. The results of such research is needed to map, as accurately as possible, the geographic range of the species and species complexes and ultimately to assist people to market artificially raised plants under the correct name. This is essential to gain and maintain the market's confidence.

Taxonomic research is also required to help train the officers who are to be responsible for the conservation and protection of the native flora.

As far back as 1972 it was felt that legislation in Papua New Guinea would discourage indescrimenent plant collecting by both professional and amature collectors. The orchids then were considered part of the national heritage (Dockrill, 1972). In fact a ban was put in place and announced on the 4th September, 1990 by Department of Environment and Conservation (Rongap, Post Courier, 5th September, 1990; Bandisch, 1990; Tanner, 1991). The ban was seen as a positive step towards the implimentation of guidelines for orchid protection and propagation as outlined by Howcroft (1990, 1991).

However, the controlling of the export of adult wild orchids from Papua New Guinea in the past has been crippled by one glaring weakness, which has been unscrupulously exploited, and that is the lack of experience and knowledge to identify the orchids at generic or species level. Orchid exporters have been found to be exporting species under false names. Without their flowers it is extremely difficult for the inexperienced to even determine that a bundle, of several orchid plants to be exported as one species, may infact contain several species. According to some informed sources plants of *Dendrobium lasianthera* have been knowingly sent out as the more common *Dendrobium discolor*. This is done to over come the legal limit placed on the numbers of plants of each species that could be exported. Orchid taxonomy can be of considerable value to such officers by enabling them to identify a 'false species' mixed with a legally exportable species.

CURRENT RESEARCH ACTIVITIES IN PAPUA NEW GUINEA

Research, in two areas, is being conducted by FRI/FCP with additional contributions from colleagues both within and outside Papua New Guinea. The first aims at determining the distribution of genera and species and the second is taxonomic research to identify genera and species.

Orchid genera and species distribution

The FRI/Flora Conservation Project is presently collecting and recording species localities from National herbarium specimen sheet records which will be used to create distribution maps of each species. additional contributions have come from Wau Ecology institute and from the Papua New Guinea orchid society.

A survey of Papuasian orchid genera associated with major New Guinea environments was recently completed (Howcroft, in press). This survey records 140 genera for Papuasia (Table 1).

Table 1. Papuasian orchid genera and their association with major environmental zones

Major Environmental Zone and Maximum Elevation	Number of Orchid Genera
Coastal Vegetation (2 m)	33
Mangroves	4
Anthropogenic Grasslands	8
Lowland Freshwater Swamps (40 m)	8
Savannah (90 m)	7
Monsoon Forest (90 m)	23
Lowland Tropical Rain Forest (700 m)	105
Lower Montane Zone (1500 m)	109
Mid-Montane Zone (2500 m)	77
Upper Montane Zone (3000 m)	48
Sub-Alpine Zone (3500 m)	27
Alpine Zone	12
Total of Genera	140

This is 6 more than previously reorded by Dockrill (1972) and Millar (1978). It identified the lowland forest and lower montane zones as having the largest number of orchid genera (105 and 109 respectively). These are zones with the greatest diversity of habitats and where forestry and farming activities occur or could occur. The figures are intended to give a broad overview, and although not sensitive enough to pin point specific areas and genera that might be threatened, it may help to help to identify some environments where certain genera exist and where investigation into the conservation status of their species before investigating others which have a far wider distribution range.

The number of genera identified in this survey is expected to increase as new taxonomic revisions of genera are accounted for and have been accepted.

A field survey of the orchids on the Woodlark islands was made in late Januaruy and early February this year in a joint FRI/FSP(Foundation for the People of the south Pacific).

Previously only 6 genera had been recorded for the islands the survey found an additional 25 increasing the recorded number to 26 genera. In addition to this 58 species were recorded whereas previously only around 19 specimens had been collected and nine identified. At least 13 species recorded are also found in North Queensland, one of which is a complete new record for Papua New Guinea.

Taxonomic Research

The Lae National Botanic Gardens staff, with FCP leader Mr. M. Kabaru, have started field collecting trips to bring specimens of native orchids back to the garden where they are to be put on display. The more desireable species will be used for breeding and propagation. Since 1992 the FCP staff have collected from

around 15 locations throughout Papuua New Guinea. Additional collections have been lodged in the Gardens by field staff from other sections. At least 4 new records have been made for Papua New Guinea from these collections and more can be expected.

Many of the species require further taxonomic treatment and this work has been initiated already. The results will be published in what is hoped will be an ongoing series of orchid research reports, the first of which was published at the end of 1992 (Howcroft, 1992). The second has been completed recently in draft form.

FUTURE NEEDS AND DIRECTIONS

Many herbarium collections that have come to hand for identification recently from sources other than FRI/FCP have been

without flowers. Identification at generic is possible with these. Whilst these are still useful records on the distribution of genera, succh collections would make a more signifiant contribution to our knowledge of species distribution if they accompanied with flowers.

The Flora Conservation Project at present concentrates on the wild orchids of Papua New Guinea. Many of these are known to also occur in Irian Jaya and recent collections have added to this list. To develop a complete picture of the overall distribution of the Papuasian orchid species there is a need to link up with other Botanists across the Papua New Guinea/Irian Jaya border to share information on the distribution of their wild orchids.

The compilation of distribution data from herbarium records may extend into 1994. A number of systems for mapping plant distributions are known to exist. One of these systems will be selected for use by the project.

Currently there is some evidence that wild orchid plants are still being taken from the country and the indications are that the 1990 ban on wild orchid export is not being taken seriously. It is believed that the Flora Conservation Project and Department of Environment and Conservation should in the near future establish better links so that the ban can be enforced until such times as the project has enough information to recommend which wild species need protection and those which can be harvested for export.

REFERENCES

Bandisch, W.H. (1990). Papua New Guinea bans export of wild collected orchids. *Newsletter of the Papua New Guinea Orchid Society.* October issue p. 2.

Cribb, P.J. (1991). The orchids of the Southwest Pacific Islands and their conservation. *Newsletter of the Papua New Guinea Orchid Society.* March issue pp 4-7.

Dockrill, A.W. (1972). Orchids. *Enyclopaedia of Papua New Guinea* (Ed. by P. Ryan), 2:870, Melbourne University Press.

Howcroft, N.H.S. (1990). Implimentations of proposed guidlines for orchid protection and propagation in Papua New Guinea. *Newsletters of the Papua New Guinea Orchid Society.* October issue pp. 13-19.

Howcroft, N.H.S. (1992). Orchid collections and research 1992 for the FRI Botanic Gardens. *Orchid Research Report* No 1. FRI Lae.

Proceedings of the Biological Society of New Guinea
R. Höft (ed.)

Putting Science into the Art of Birdwing Butterfly Ranching and Conservation: Invoking Basic Ecological Concepts

Larry Orsak

Christensen Research Institute, P.O. Box 305, Madang, Papua New Guinea

ABSTRACT

The literature on butterfly ranching in Papua New Guinea, as well as conservation strategies for Papua New Guinea's unique birdwing butterflies, which include the world's largest and second largest species, are largely in the form of unpublished reports. Unfortunately, the lack of refereed evaluations of these reports have resulted in many unsubstantiated statements, which nevertheless have been repeatedly cited. The world's largest butterfly, Queen Alexandra's Birdwing (*Ornithoptera alexandrae*), has been an especially common victim of this process. The butterfly has been tested as a remarkably unique species that is exceptionally rare, and whose males must feed at Kwila flowers in order to mate, and whose females look forthe prior presence of caterpillars in order to determine which plants they themselves will lay eggs on.

The big problem with these and other suppositions is that they are then used to formulate management and butterfly ranching strategies. Although some of these theories will need to be experimentally tested in order to determine their validity, most of them can simply be compared with what is known about butterflies in general, or held against widely accepted ecological theories, such as niche partitioning and density-dependent regulation of populations, to see that they are almost certainly not true. In this paper, I go over statements that have been made about birdwing butterflies in Papua New Guinea, compare them against theory and previous research on butterflies, and indicate where the serious flaws in developing butterfly ranching and conservation strategies lie.

Proceedings of the Biological Society of New Guinea
R. Höft (ed.)

USE AND CONSERVATION OF PAPUA NEW GUINEA'S FORESTS

Bas T. Louman

Department of Forestry, Papua New Guinea University of Technology, Private Mail Bag, Lae, Morobe Province, Papua New Guinea

ABSTRACT

The natural forests in Papua New Guinea can perform many functions to support human, animal and plant life. This paper discusses how multiple forest use planning should enable Papua New Guineans to manage their forest resources in such a way that all of the forest functions can be met more or less simultaneously and to an extend that is satisfactory to as many interest groups as feasible. Management should be based on the principles of sustainable forest use, conservation of the forest as ecosystem, and preservation of specific forest types.

Within such management systems there is a place for both commercial timber harvesting and pure protection, depending on the needs of the local population, the country, and the environment. Timber harvesting may also be an option for conservation oriented management systems. In that case, strict rules will have to be adhered to, and harvesting should form an integrated part of an overall land use plan.

Sustainable management systems in Papua New Guinea should take into consideration the specific nature of land and forest ownership in Papua New Guinea. This will usually mean that sustainability should be strived for on a project basis, and even management unit basis. Optimum size of such management units for harvesting operations have been found to be around 30,000 ha in the rain forest of Surinam (South America), but may have to be slightly bigger in Papua New Guinea. Smaller sizes of timber operations will also better serve biodiversity purposes. An example is given based on information from the proposed April-Salumei timber area.

INTRODUCTION

Papua New Guinea has a very active forest industry, producing annually as much as two million cubic meters of export logs at a value of 165 million Kina (Department of Forests 1992), and approximately 250,000 cubic meters for local use (Department of Forests 1991). Until 1992 this resulted in an area of up to 100,000 ha tropical rain forest being logged annually. At present an estimated 1 million hectares of forest have been logged over, another 2-3 million hectares are on the verge of being logged.

At the same time, agriculture claims annually approximately 20,000 ha of virgin forest (Department of Forests 1991, IUCN 1991), while every year approximately 200,000 ha of forest land, principally secondary forest, is affected by shifting cultivation (IUCN, 1991).

Forests in Papua New Guinea also provide most of the population with essential elements for subsistence (fruits, protein, timber, leaves for thatched roofs, medicines, etc) and some commercially harvestable non-timber products such as copal, rattan, massoy, cassowary, wild pigs, and insects such as butterflies. An estimated 5 million cubic meters of wood is annually chopped up into firewood (World Resources Institute 1991).

Finally, but not least, the forest provides protective functions for people in and outside the forest, provides a habitat for thousands of plant and animal species, and plays a major role in maintaining the present living conditions on both local and global scale (e.g IUCN 1991, Tisseverasinghe 1988).

The paramount question is, therefore, can we manage the forest resources in such a way that all of these forest functions can be met to a satisfactory extend, both at present as well as in the future, and if so, how this can be done.

The answers to these questions are probably as manifold as there are academics and politicians, and it would be presumptuous of me to pretend to come with a solution that is preferable above any of the solutions thought out previously by other people. I do contend, though, that this paper provides a more objective approach to the different aspects of management of the forest for multiple uses than has been shown over the past few years by ecological activists, logging companies, forest owners, and the different politicians interested in forestry and conservation matters.

First of all, there is no such thing as the only best or most important usage of the forest. Any of the above mentioned functions of forests may assume primary importance, depending on the local situation, the availability of alternatives, and the possibility to receive the benefits from the forests elsewhere. A review of management of the natural forest in India, Malaysia, and the Philippines came to similar conclusions: any choice for forestry as the major type of land use in an area (including conservation and protection), should be based on objectives derived from the social context, but within the limits of technical possibilities (FAO 1989).

Any policy advocating a blanket approach to Papua New Guinea's forests is, therefore, an erroneous policy. While national policies and plans designating areas for specific types of land use are necessary, such policies and plans should be flexible enough to allow for considerable local adjustments. National policies and plans should have coordinating roles, rather than form directives.

This is in particular the case in Papua New Guinea, since here very little is known about different types of land use and their sustainability under different circumstances.

Papua New Guinea is a conglomerate of private entities (land and forest owners), each of which will have to be satisfied in order to come to sustainable development. Part of this satisfaction is derived from the ability to have a say in planning the development of one's own resources. Part, also, can be derived from participating in both costs and benefits of all of the different forest resources.

Thus, it is not feasible under present circumstances in Papua New Guinea to reserve large areas for only conservation purposes, nor should large areas be reserved for the production of one single forest product. Instead, multiple use plans should be drawn up for each Forest Management Area, each Conservation Area, each agricultural area, etc., where the principal uses can be as indicated by the respective national plans, while simultaneously the areas are providing one or more of the other functions as well.

For all of the different functions that forest may directly, or indirectly, provide, three principles should prevail: 1) The use of the function has to be sustainable, in such a way that future generations can continue to benefit from the same functions, 2) to be able to do that, some type of forest ecosystem will have to be maintained in sufficiently large areas, and 3), enough primary forest area should be preserved to maintain open the future options for use of specific products/functions.

The Three Principles and Logging

The three principles mentioned in the previous paragraph strongly emphasise sustainability, maintaining forest ecosystems, and preserving primary forests. At the same time, tree harvesting for commercial purposes is, and has been, one of the principal uses of the natural forest. Should Papua New Guinea continue to log the natural forest in order to earn extra revenue, or will this be contrary to the stated principles? The answer to this question is simple: Yes......and, No.

Logging and sustainability

Why these seemingly contradictory answers? Because logging, as it is practiced in Papua New Guinea at present can definitely not be considered to be compatible with any type of sustainable use of the forest. The approach towards logging operations is of too large a scale to continuously involve the same local groups of people, terms of planning are of too short a duration for any future problems to have an impact on the present activities, in relation to that, the access to resources is too insecure to be able to plan for any longer terms (say 20 years and beyond), and, with some exceptions, companies consider logging to be purely an activity to be used to extract timber from the forest at minimum cost and maximum profit, rather than as part of a (silvicultural) system that in its turn supports the management of the forest.

Even ecologically supposedly benign harvesting operations such as helicopter logging will not be sustainable under such conditions. You cannot expect to harvest 200,000 ha within a period of three years, then follow on to the next 200,000 ha, and at the same time leave the local population satisfied with their three year involvement. Satisfied enough, that is, to independently protect and maintain their remnant forests for another 25-40 years until the helicopters will be back for the next felling cycle.

While the revenue to the local people may have been great during those three years, most of the villages will not have had the capacity to use that revenue to set up any other type of sustainable, revenue earning activity. Expectations of higher living standards will have been created, and when these do not come through, except for possibly the first four to five years after the commencement of logging, then the people will be disillusioned, and socially and economically worse off than before harvesting started. Worse even, if planning of the operations is of short to medium term nature, and there is no guarantee that the company will be back after those 25-40 years. Under those circumstances the remaining forest is of little value to the land owners, and the risk is great that the forest will have to make place for other economic activities.

This does not mean, however, that logging per definition cannot be compatible with conservation and protection of the rain forest. A good example is the experience obtained in the Queensland tropical rain forests. In 1988 900,000 ha of tropical rain forest were declared World Heritage Area. This was only possible, because during the previous 15-20 years logging activities gave an explicit value to the forest, and therefore made maintenance of the forest ecosystem compatible to other types of land uses.

As little as 60,000 cubic meters per year were harvested from an area of more or less 900,000 ha, though harvests on a per hectare basis were in the order of 30 cum. All harvested timber was locally processed. Approximately 1,800 people were directly employed in the managing, growing and harvesting wood, while another 4800 were employed in the processing industry (Resource Assessment Commission, 1992). At the same time, logging activities had very little long term impact on the ecology of the rain forests (Oldeman and Van der Meer 1988).

Logging and conservation

The above example shows that timber harvesting operations can be very conservative. An essential feature of the operations in Queensland is the planning and organisation of the activities, both administrative as well as in the field. Based on detailed inventories and information from Permanent Sample Plots, the major activities are planned on paper. These entail the area that should be logged, the amount of timber that may be harvested from that area in a specified period of time, the method of harvesting that should be employed in order to safeguard regeneration, and the length between two subsequent cuts.

With that information, planning is taken into the field: plans of roads, bridges, culverts, ramps, and skidding trails are drawn up and alignment takes place in the field. All trees to be cut will be marked, and so will trees that have to be retained either for future harvests, or as seed trees. Marking includes indicating the felling direction.

· Skilled chainsaw operators closely follow the felling rules, after which the tree markers return to monitor the damage done to the forest. Enforcement of the rules is strict.

Subsequently, controlled skidding takes the trees out of the forest in log-form, and where necessary skidding trails are ripped, riverbeds cleared of logging/skidding debris, and/or sections of trails and roads completely closed.

The need for well planned, strictly implemented and closely monitored operations has also been stressed in the previously cited FAO publication, reviewing natural forest management systems in Southeast Asia (FAO 1989). In Southeast Asia, suggests the review, much of the failures of such management systems is due exactly to the lack of strict implementation and monitoring. Lack of sufficiently qualified and experienced manpower, together with insufficient incentives for the few qualified people to remain in the field, may be the principle causes for poor implementation of otherwise promising forest management systems.

At present, Papua New Guinea too lacks the skills to properly plan, execute, and monitor these operations. In addition there appears to be a lack of motivation to improve present logging activities at local village level, at company level, as well as at the political level. The principle cause for this seems to be the fear of loss of revenue.

However, on the long term more revenue will be lost if present operations do not improve soon: damage to the forest will become so severe that Papua New Guinea can no longer afford to continue commercial logging of its forests, and a complete ban may have to be enforced, not in the least due to international pressure. At the same time local people badly in need of some type of income will start their own cut and sell practices over which exists even less control, causing further forest degradation. These practices can already be found around Lae.

In order to maintain the forest revenues at a reasonable level, it will be necessary to start to limit the logging practices to those approaching sustainable harvesting practices. Simultaneously a much greater effort will have to be done to develop for Papua New Guinea relevant forest management systems, which address maintenance of all the possible forest functions.

In addition, all those areas that at present cannot be logged on a sustainable basis, and for which no direct other type of sustainable land use (including conservation and preservation) is planned, should be declared forest reserves. These reserves are to be set aside for future forest use, and should therefore be protected at present. When more information on these areas will become available, more experience has been obtained with the different types of forest use, and local people have gained in awareness and education, a more definite use of these forest areas can be determined.

Logging and preservation

In this context I refer to preserving as maintaining something (the forest) in it's present state. As such, logging of a particular forest stand is definitely not compatible with preservation of that same stand.

On the other hand, we have to ask ourselves, how much of the present rain forest should really be preserved in its present state? This depends on the objective of preservation. In general, though, that objective is directed at maintaining genetic diversity, both within the species and between species. It is, therefore, not sufficient to protect specific species. Samples of a large variety of different ecosystems will have to be sustained to reach the above objective of preservation. Each of these samples may be of rather limited size. Zobel and Talbert (1984) have suggested that for tree species a minimum area of 2000 ha has to be maintained. For some animal species it may have to be much more than that, but for many animal species the minimum habitat size that needs to be protected in order to maintain the species is not yet known. Lawrence (1993, pers. comm.) indicates that for some species it is sufficient to maintain small areas that are interconnected by gallery forests.

Many species will survive under controlled harvesting conditions, some even due to human interference. Specifically vulnerable species, however, as well as specific interactions between some components of the ecosystem will be affected by any type of human interference.

PLANNING OF A MULTIPLE USE FOREST MANAGEMENT AREA

The proposed April Salumei timber area in the East-Sepik province of Papua New Guinea (Figure 1) is an interesting example of an area that is expected to yield different benefits for different interest groups. The area is large (500,000 ha) and relatively un-inhabited (less than 1 person/km^2).

Figure 1. Sketchmap of the proposed April-Salumei Timber area in the East Sepik Province

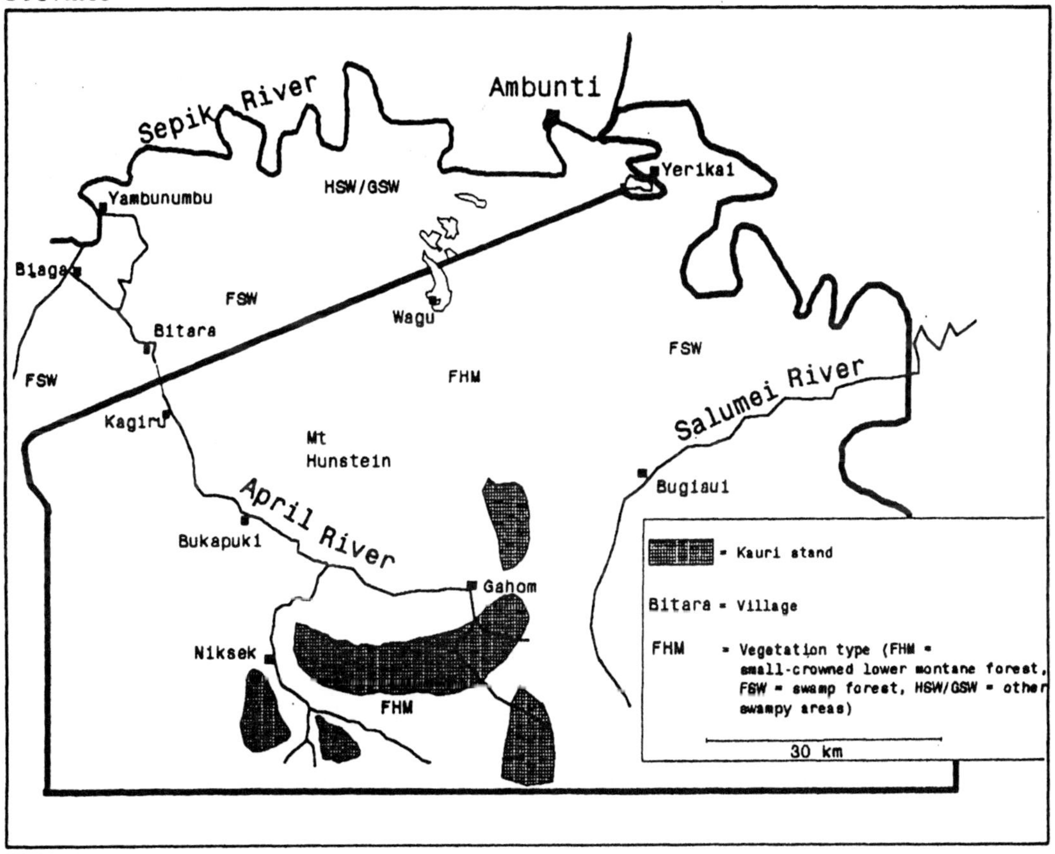

The area has been ear-marked as having a very high priority for bio-diversity conservation (Department of Environment and Conservation 1993, Lawong this meeting), with an interesting, but still largely unknown, fauna and flora. In the recent past exploration for minerals took place in the area with mixed results. At the moment both the provincial and national governments have a major interest in developing the area through timber extraction (the area has some commercially very interesting kauri (*Agathis* sp.) stands), while local groups are more interested in small-scale exploitation of forest products through the use of portable sawmills, for production of artefacts and canoes, or for direct sale (rattan, copal, wild life)(Unitech Development and Consultancy 1993, FOTP 1991).

In planning the use of this particular area, one should first of all make an analysis of what may happen to the area if no such planning would take place.

Under the present political and socio-economic climate in Papua New Guinea it is very likely that within not too long a time the area will be allocated to any logging firm that promises to establish the necessary social and physical infrastructure, and intends to start sawmilling the local timber after 3-5 years (common procedure in recent concession areas).

While present forestry policy and legislation are geared towards directing such enterprises towards sustainable utilisation of the forest, neither are without loop holes, nor is any legislation any better than it's implementation in the field. Unfortunately, Papua New Guinea has a very poor record in this respect. This is partly due to the vast forest resources, which make it difficult to control everything that is going on in and around them. Partly, also, it is due to insufficient manpower, finance and commitment of the former Forestry Department to be able to simultaneously monitor all forestry operations throughout the country.

In particular in remote areas, and in the case of large forest concessions, it becomes very difficult to steer operations in the right direction, unless that direction would benefit all participants (i.e. landowners, forest companies, government). This, however, is seldom the case.

Steering of forest operations in a direction that is acceptable to all active and passive forestry participants, will require a very different approach from the conventional forest concessions, starting at the planning level, and going all the way through to participation in the actual utilisation of the forest resources and in the distribution of both costs and benefits derived from such participation.

The main role of the government in such an approach, will be to safeguard the rights of society as a whole, and specific groups in the society affected by the activities in particular. For example, in the case of timber extraction activities along the April river, the people living downstream along both the April river and the Sepik river will almost certainly be affected, either by water pollution as a result of the actual extraction activities, or by pollution or disturbances caused by related activities, such as transport of employees and materials by motorised canoes or using a road to be constructed.

The government will have to ensure that such negative affects on neighbouring population groups remain within reasonable limits. It can do so through enforcing specific policies and laws. For that, it will need a presence in the area. Not only should the government representative have the authority and respect to deal with offences in a decisive manner. He (or she) should also have a catalytic role in the planning and coordination of the utilisation of the forest resources.

Decisions regarding (forest) land use, however, should be taken by all participating groups, as long as they remain within the framework (policy & legislation) set by the national governing bodies (*i.e.* National Forest Service). Implementation of those decisions should be by those people that are best equipped to do so.

In my opinion, management of forestry concessions should be based on a project basis, with a professional forester as project manager who shall seek the advice in all policy and planning matters of a project board/authority. The board should consist of a government representative (NFS), landowners/local governments, and one representative for each of the potential commercial activities (e.g. large scale logging, small-scale logging, small-scale mining, agriculture, fishing, artefact manufacturing). It's advice will not be binding, but any deviation from it's advice will have to be justified by the manager, and his decisions can be appealed against with his employer, the National Forest Service.

The manager should have a supportive staff existing of some staff positions (Rural development specialist, Agricultural specialist), administrative positions (including a finance and loans officer running a rotational fund to borrow money to local people wanting to set up small projects), and technical positions that will handle the forest management and conservation aspects of the project (the Forestry Specialist, Forest Management Team, Conservation and Protection Team, and the Extension Worker(s) in Figure 2).

The principle role of the staff will be to support the planning process, to support local developments, to run the administrative matters of the project, and to supervise all participants in the project with regard to their adherence to project agreements.

Figure 2. Proposed management structure April-Salumei project (adjusted from UDC 1993 p. 53)

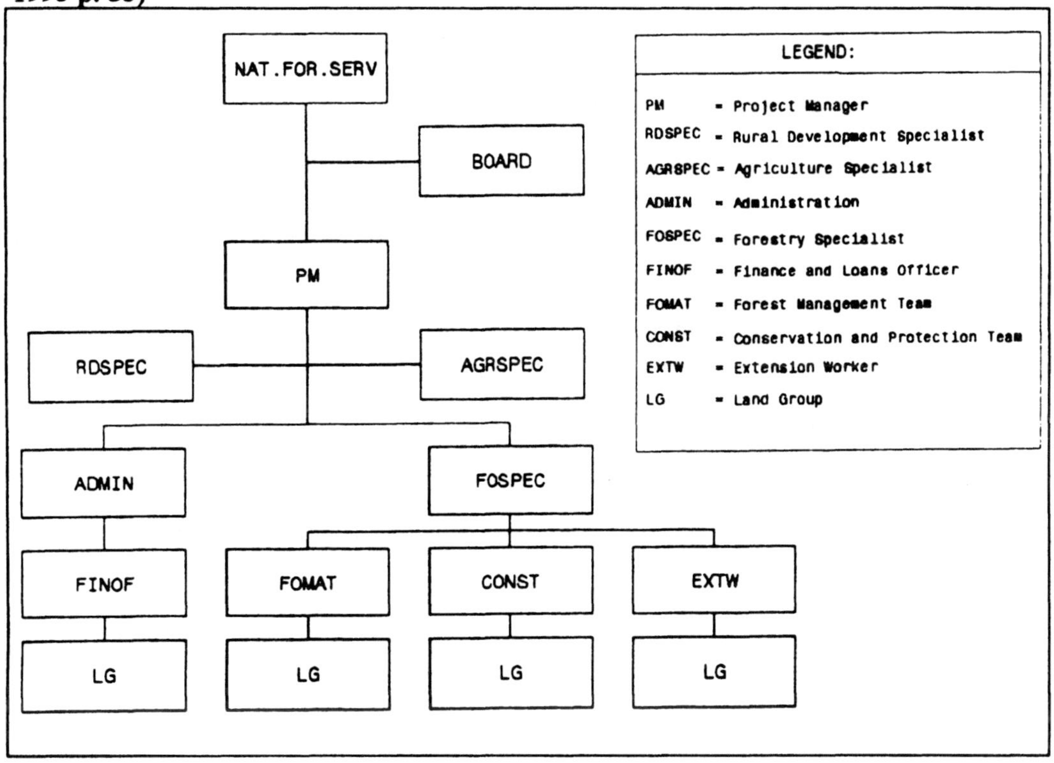

The project should receive its income partly from any companies operating in it's area (in the form of levies), partly from the government (from the government's share of royalties), and partly from the local people (from the funds held in trust by the government).

Within such a project the first activity has to be the formulation of a land use plan, taking into consideration the desires of the interest groups as they are represented in the project board. Assuming that in our example the area has been proposed as a timber area, the predominant use of the area should be forestry oriented.

At this stage, it is good to consider some forest management experiences from Surinam (South America), where it was found that optimum sustained yield forest management units had a size of 20,000-30,000 ha (Hendrison, 1990). Such a unit proved to be easily manageable, both from the administrative, as well from the field technical point of view. One supervisor and a logging unit of one tractor skidder, two wheeled skidders, and three chainsaw crews were able to cope with all the necessary harvesting operations, while a small crew of locally recruited labourers dealt with the necessary silvicultural operations.

In Papua New Guinea, a sustained management unit would have a minimum area slightly bigger than in Surinam, since the proposed cutting cycle is not 25 years, as in Surinam, but 40 years. Thus areas of 40,000-50,000 has appear to be appropriate.

Such sizes would also meet the requirement of forestry operations to try to achieve a more human scale. Admitted, 40,000-50,000 ha is still considerably more than what most clans own in Papua New Guinea, but the number of clans involved in such an area will be limited, making a common approach towards the management of the area easier.

Going back to our example, this would mean that the 500,000 hectares of the April-Salumei area could be split up into 10 to 12 management units, each of which can be managed for the production of a sustained yield. Taking the average commercial volume of forests in Papua New Guinea of 30 m^3 (IUCN 1991) each such unit would produce 30,000 to 37,500 m^3

of timber from 1,000 to 1,250 ha each year. At average FOB prices of K 200/m^3 this timber would have a value of approximately 6-7.5 million Kina in it's unprocessed form.

This timber quantity is in principle sufficient to maintain a medium sized sawmill operation. However, the remoteness of the area makes it necessary to substantially invest in the infrastructure necessary to communicate with the outside world. Figures of up to K 3 million for road construction have been mentioned (Unitech Development and Consultancy 1993). Such amounts cannot be born by small to medium sized industries.

Some time in the near future, a road will be necessary to support other types of infrastructure and the desired services in the area.

It appears therefore desirable to find a means to finance such a road. One way would be to contract a genuine forest investor, interested in operating a larger scale local processing industry. This could, for example, be an investor that is prepared to establish a plywood mill. A small to medium sized mill would need an overall annual input of approximately 50,000 m^3 of top-quality timber to be viable (Unitech Development and Consultancy 1993). This is equivalent to two to three of the previously mentioned management units, depending on the availability of quality timber (Kauri-pine) in each unit.

A medium sized plywood mill would be able to assist in the construction of a road, preferably by paying the costs of contracting a road construction firm. It would also assist local development by providing direct employment in the mill and in the field. It would further provide indirect employment by generating a wage-earning group in the population which will desire to buy items such as canoes and food. Wage-earners in general attract services such as aid-posts, government services, schools, and shops.

The road and the services combined will facilitate the development of local small businesses, that is, if in addition local people are supported in such developments by appropriate organisations (e.g. ESCOW, in our example: the project management).

What is more, since the mill would only depend on two to three selected management units, the rest of the April-Salumei area (350,000-400,000 ha) can be managed to achieve different objectives.

One could imagine one or two forest management units to be logged by contractors for the log-export or local sawmilling. One could also imagine to clear small areas for agriculture and forest plantations. Areas could be reserved for hunting, collecting etc.

More importantly, though, a large part of the area can be set aside for conservation and protection. A conservation and protection that will receive substantial benefit from the planned economic activities in the project area: in the form of financial support, but also because the other activities, if well planned and properly monitored, will be able to provide most of the needs of the local population.

CONCLUSION

Papua New Guinea is fortunate to have a relatively small area deforested every year. The pressure on the natural forest is mounting, though, and a need exists to safeguard the existing forests from excessive damage and destruction.

Conservation and protection of large areas of land is one way of doing this. Experience elsewhere, however, has shown that if conservation and protection projects do not simultaneously meet the need of local people to receive services and increase their welfare, they are bound to fail. In order to meet those needs, multiple land use projects have started to develop in many tropical countries.

The idea of multiple land use projects is not new to Papua New Guinea (think for example of the concept of Integrated Conservation And Development (ICAD)). What is new in this proposal is the integration of large scale logging and processing activities in such an ICAD.

Through an example of a proposed timber area, it is suggested that such larger scale activities will contribute to the financial viability of the project and will facilitate the generation of small scale economic activities. However, this is feasible only if the project is under professional management, preferably headed by a forester, that is monitored by a local advisory board including all the forestry participants relevant for the area.

REFERENCES

Department of Environment and Conservation (1993). *Papua New Guinea Conservation Needs Assessment Synopsis Report.* Department of Environment and Conservation, Boroko.

Department of Forests (1991). *Country Report 1991.* Department of Forests, Hohola.

Department of Forests (1992). *Timber Digest December 1992.* Department of Forests, Hohola.

FAO (1989). *Review of forest management systems of tropical Asia. Case studies of natural forest management for timber production in India, Malaysia and the Philippines.* FAO Forestry Paper 89, Rome.

FOTP (Friends of the Sepik) (1991). *The Hunstein Range, integrating enterprise, culture and conservation in the Upper Sepik, Papua New Guinea.* FOTP Wewak/London.

Hendrison, J. (1990). *Damage-controlled logging in managed tropical rain forest in Suriname.* Agricultural University Wageningen, The Netherlands.

IUCN (1991). *The Conservation Atlas of Tropical Forests. Asia and the Pacific.* International Union for the Conservation of Nature, MacMillan Press, London.

Oldeman R.A.A., & Van der Meer, P.J. (1988). *Diagnosis of Northern Queensland rain forests, the impact of selection silviculture.* Agricultural University of Wageningen, Netherlands.

Resource Assessment Commission (1992). *Forest and Timber Inquiry, Final Report Vol 1.* Commonwealth of Australia, Canberra.

Tisseverasinghe, K. (1988). *Introduction to forestry.* University of Technology, Lae.

Unitech Development and Consultancy (1993). Feasibility study April-Salumei Area, Sustainable development based on forest resources. Report prepared for the East Sepik Province Division of Forests. East Sepik Province Division of Forests, Wewak.

World Resources Institute (1990). *World Resources 1990-91. A guide to the global environment.* Oxford University Press, New York.

Zobel, B. & Talbert, J. (1984). *Applied forest tree improvement.* John Wiley & Sons, Inc. New York.

CPSIA information can be obtained at www.ICGtesting.com
Printed in the USA
LVOW030136200712

290832LV00001B/1/P

9 789980 945938